Production Management, Manufacturing, and Process Control

Drawing on contributions from various manufacturing fields, this book offers a comprehensive perspective by combining theoretical concepts with practical applications. It emphasizes future developments, the integration of technologies, and the crucial role of humans in manufacturing companies.

Production Management, Manufacturing, and Process Control presents cutting-edge strategies and innovations for creating people-centered manufacturing processes. It explores how culture influences cognition and behavior, providing readers with valuable insights into relevant theories. This book also explores risk management, human performance improvement, and the current challenges in quality and information systems management. Sustainable global manufacturing practices that balance global market access with strong domestic engineering ecosystems are covered in detail, and this book also addresses the optimization of production processes, including the use of machine learning for fault diagnosis.

This is an ideal read and a valuable resource for students, graduates, teachers, researchers, and professionals in industrial management, business management, safety fields, manufacturing, risk management, and quality management.

Industrial and Systems Engineering Series

Series Editor: Waldemar Karwowski, University of Central Florida, Orlando, USA & Hamid Parsaei, Texas A&M University

Industrial Engineering has evolved as a major engineering and management discipline, the effective utilization of which has contributed to our increased standard of living through increased productivity, quality of work, and quality of services and improvements in the working environments. The Industrial and Systems Engineering book series provides timely and useful methodologies for achieving increased productivity and quality, competitiveness, globalization of business and for increasing the quality of working life in manufacturing and service industries. This book series should be of value to all industrial engineers and managers, whether they are in profit motivated operations or in other nonprofit fields of activity.

Design and Construction of an RFID-enabled Infrastructure: The Next Avatar of the Internet
Nagabhushana Prabhu

Managing Professional Service Delivery: 9 Rules for Success
Barry M. Mundt, Francis J. Smith, and Stephen D. Egan Jr.

Mobile Electronic Commerce: Foundations, Development, and Applications
Edited by June Wei

Manufacturing Productivity in China: Edited by Li Zheng, Simin Huang, and Zhihai Zhang

Human Factors in Transportation: Social and Technological Evolution Across Maritime, Road, Rail, and Aviation Domains
Giuseppe Di Bucchianico, Andrea Vallicelli, Neville A. Stanton, and Steven J. Landry

Laser and Photonic Systems: Design and Integration
Edited by Shimon Y. Nof, Andrew M. Weiner, and Gary J. Cheng

Supply Chain Management and Logistics: Innovative Strategies and Practical Solutions
Edited by Zhe Liang, Wanpracha Art Chaovalitwongse, and Leyuan Shi

Production Management, Manufacturing, and Process Control
Edited by Beata Mrugalska, Waldemar Karwowski, and Tareq Z. Ahram

For more information about this series, please visit: https://www.routledge.com/Industrial-and-Systems-Engineering-Series/book-series/CRCINDSYSENG

Production Management, Manufacturing, and Process Control

Edited by

Beata Mrugalska, Waldemar Karwowski, and Tareq Z. Ahram

NEW YORK AND LONDON

Designed cover image: Shutterstock - Gorodenkoff

First published 2025
by Routledge
605 Third Avenue, New York, NY 10158

and by Routledge
4 Park Square, Milton Park, Abingdon, Oxon, OX14 4RN

Routledge is an imprint of the Taylor & Francis Group, an informa business

Library of Congress Cataloging-in-Publication Data
Names: Mrugalska, Beata, author, editor. | Karwowski, Waldemar, 1953-
editor. | Ahram, Tareq Z., editor.
Title: Production management, manufacturing, and process control / edited
by Beata Mrugalska, Waldemar Karwowski, Tareq Z. Ahram.
Description: First published. | New York, NY : Routledge, 2025. |
Series: Industrial and systems engineering series | Includes bibliographical
references and index.
Identifiers: LCCN 2024014538 (print) | LCCN 2024014539 (ebook) |
ISBN 9781032825977 (hardback) | ISBN 9781032825984 (paperback) |
ISBN 9781003505327 (ebook)
Subjects: LCSH: Manufacturing processes—Technological innovations. |
Manufacturing processes—Human factor.
Classification: LCC TS183 .P7525 2025 (print) | LCC TS183 (ebook) |
DDC 670—dc23/eng/20240806
LC record available at https://lccn.loc.gov/2024014538
LC ebook record available at https://lccn.loc.gov/2024014539

ISBN: 978-1-032-82597-7 (hbk)
ISBN: 978-1-032-82598-4 (pbk)
ISBN: 978-1-003-50532-7 (ebk)

DOI: 10.1201/9781003505327

Typeset in Times
by codeMantra

Contents

Preface

This book discusses the latest advances in a broadly defined field of advanced manufacturing and process control. It provides information on the latest strategies and innovations for people-centered manufacturing processes. It shows the impact of culture on people's cognition and behavior, providing readers with timely insights into theories. Moreover, it focuses on risk management and human performance improvement. It presents a comprehensive overview of the current challenges in the management of quality and information systems. This book also sheds light on sustainable global manufacturing practices balancing global market access with robust domestic engineering ecosystems. It also analyzes the issues of optimization of production processes including machine learning for fault diagnosis.

Drawing on contributions from the authors within a variety of manufacturing fields, it presents a broad perspective. It brings together theoretical and practical practices highlighting the future developments and integration of technologies and human role in manufacturing companies.

Beata Mrugalska
Waldemar Karwowski
Tareq Ahram

About the Editors

Beata Mrugalska, Ph.D., D.Sc., Eur.Erg., is an associate professor and head of the Division of Applied Ergonomics, Institute of Safety and Quality Engineering, Faculty of Management Engineering, Poznan University of Technology in Poland. She holds an M.Sc. (2001) in Management and Marketing from the Faculty of Mechanical Engineering and Management at the Poznan University of Technology and a Ph.D. (2009) in Machine Construction and Operation from the Faculty of Computer Science and Management at the Poznan University of Technology. She was awarded a D.Sc. (dr habil.) degree in Mechanical Engineering by the Faculty of Mechanical Engineering at the Poznan University of Technology, Poland (2019). Since 2018, she has been a board member of the Center for Registration of European Ergonomists (CREE). She is responsible for promoting the professional title of EuroErgonomist in the international scientific and industrial environment. She is also a board member of the Polish Academy of Arts and Sciences, Commission of Ergonomics, and a member of the Ergonomics Committee at the Polish Academy of Sciences, Branch Poznan. She has over 120 publications focused on human factors in modern organizational management concepts. She serves as a member of eight editorial boards of international journals and a guest editor of four special issues of journals. She was honored with the Top Peer Reviewer by Publons for her contribution to the preparation of paper reviews.

Waldemar Karwowski, Ph.D., D.Sc., P.E., is a Pegasus Professor and chairman of the Department of Industrial Engineering and Management Systems and executive director of the Institute for Advanced Systems Engineering, University of Central Florida, Orlando, Florida, USA. He holds an M.S. (1978) in Production Engineering and Management from the Technical University of Wroclaw, Poland, and a Ph.D. (1982) in Industrial Engineering from Texas Tech University. He was awarded D.Sc. (dr habil.) degree in Management Science by the State Institute for Organization and Management in Industry, Poland (2004). He is past president of the International Ergonomics Association (2000–2003) and the Human Factors and Ergonomics Society (2007). He served on the Committee on Human Factors/Human Systems Integration, National Research Council, and the National Academies, USA (2007–2011). He has over 500 publications focused on human performance, safety, cognitive systems engineering, human-centered design, neuro-fuzzy systems, nonlinear dynamics, neurotechnology, and neuroergonomics. He serves as the co-editor-in-chief of *Theoretical Issues in Ergonomics Science* journal (Taylor & Francis, Ltd), editor-in-chief of *Human-Intelligent Systems Integration*, and field chief editor of *Frontiers in Neuroergonomics*.

Tareq Ahram, Ph.D., is an assistant research professor and a lead scientist at the Institute for Advanced Systems Engineering at the University of Central Florida, Orlando, Florida, USA. He received a Ph.D. degree in Industrial and Systems

Engineering from the University of Central Florida in 2008, with specialization in human systems integration and large-scale information retrieval systems optimization; M.Sc. in Industrial Systems Engineering – Human Factors from UCF in 2007; M.Sc. and B.Sc. in Industrial and Systems Engineering from the University of Jordan in 2002 and 2004, respectively. He has served as an invited speaker and a scientific member for several systems engineering, emerging technologies, neurodesign, and human factors research, and an invited speaker and a program committee member for the U.S. Department of Defence Human Systems Integration and the Human Factors Engineering Technical Advisory Group. Tareq Ahram served as an invited research professor and expert internationally and an active research member in various US- and EU-funded research projects. He serves as the applications subject editor member of *Frontiers in Neuroergonomics* and the executive editor of the *Journal of Human-Intelligent Systems Integration.* He is the recipient of the US Outstanding Researcher and the IBM Smart Systems Research Award. He is currently the lead scientist and an assistant research professor working at the Institute for Advanced Systems Engineering (IASE), USA.

Contributors

Zeki Anıl Adıguzel
Department of Industrial Engineering
Baskent University
Ankara, Turkey

Ahmed Baran Azizoglu
Department of Industrial Engineering
Baskent University
Ankara, Turkey

Adrián Morales-Casas
Institute of Biomechanics of Valencia
Valencia, Spain

Esra Dinler
Department of Industrial Engineering
Baskent University
Ankara, Turkey

Adriano Gomes de Freitas
Institute of Management and Information Systems
Monash University
Melbourne, Victoria, Australia

Lorenzo Solano-García
Institute of Design and Manufacturing
Universitat Politècnica de València
Valencia, Spain

Krzysztof Hankiewicz
Poznan University of Technology
Poznan, Poland

Janne Heilala
Department of Mechanical and Materials Engineering
University of Turku
Turku, Finland

José Laparra-Hernánez
Institute of Biomechanics of Valencia
Valencia, Spain

Yusuf Tansel Ic
Department of Industrial Engineering
Baskent University
Ankara, Turkey

Fatemeh Davoudi Kakhki
Machine Learning & Safety Analytics Lab
Department of General Engineering
Santa Clara University
Santa Clara, California

Ceren Karagöz Katı
Department of Defence Technologies and Systems
Baskent University
Ankara, Turkey

Alparslan Eren Keskin
Department of Industrial Engineering
Baskent University
Ankara, Turkey

Minesu Koksal
Department of Industrial Engineering
Baskent University
Ankara, Turkey

Paweł Królas
Institute of Management and Information Systems
Poznan University of Technology
Poznan, Poland

Armin Moghadam
Department of Technology
San Jose State University
San Jose, California

Beata Mrugalska
Poznan University of Technology
Poznan, Poland

Berat Subutay Ozbek
Department of Industrial Engineering
Baskent University
Ankara, Turkey

Susan Postlethwaite
MFI
Manchester Metropolitan University
Manchester, United Kingdom

Kat Thiel
MFI
Manchester Metropolitan University
Manchester, United Kingdom

Keiko Toya
Meiji University Tokyo
Tokyo, Japan

Amparo López-Vicente
Institute of Biomechanics of Valencia
Valencia, Spain

1 Creativity of Employee 4.0

Individual and Collective Roles of Personal and Contextual Factors

Beata Mrugalska

1.1 INTRODUCTION

Apart from moving many companies to transition to and engage in online activities, a digital revolution also brings various opportunities, including new possibilities of connecting and exchanging data. It combines emerging technologies such as Artificial Intelligence (AI), Internet of Things (IoT), Big Data, and Machine Learning (ML). These technologies enable teams, production lines, and business process to work together not taking into account distance, time zone, network, or any other factor. They facilitate the integration of user experience and lean manufacturing (Yousefi et al., 2020). The use of cloud storage enables linking products to companies on a global scale and management of the extensive volumes of big data generated by IoT. Furthermore, it serves as a widely accepted platform, which enables its resources accessible to users remotely over a network without human-to-human or human-to-computer interactions. Along with the creation of new goods and services, it can result in better decision-making and predictive maintenance. Generally, cloud computing can offer the platform and infrastructure required to enable Industry 4.0's cutting-edge technologies and data-driven methods, resulting in a manufacturing process that is more inventive, adaptable, and efficient (Mrugalska & Stasiuk-Piekarska, 2020; Mrugalska & Ahmed, 2021; Ahmed et al., 2022).

However, the importance of the impact of Industry 4.0 is mainly directed on workers as it leads to the enhancement of efficiency in enterprise. The role of human capital, which is nowadays the primary asset for any organization and whose effective utilization enables accomplishing the enterprise's objectives, is also strongly visible in fulfilling present and future market demands and gaining a competitive advantage (Sollosy, 2016). Therefore, the era of Industry 4.0 is strictly connected to adjustments in the education system of prospective employees, which has to be focused on future expected competencies and skills. The changes will also be observed in the labor market and human resource management (HRM) practices (Piwowaj-Sulej, 2020; Umair et al., 2023a, b, c). It will be particularly noticed when we consider the

DOI: 10.1201/9781003505327-1

employee's role in a company and the changes in reference to qualifications and professional competencies, particularly digital skills and the ongoing process of professional development (Ejsmont, 2021). Moreover, in the Fourth Industrial Revolution, organization's main objectives are oriented toward innovation and improved performance. In order to promote these goals, we need creative employees whose imaginativeness will correlate with productivity and who will use all available resources to achieve specific objectives. Therefore, this chapter is devoted to the problem of idea generation and knowledge structures. Personal and contextual factors are analyzed as they are helpful in creating occupational profiles for workers in organizations that plan to take a direction toward Industry 4.0.

1.2 IDEA GENERATION

The generation of unique and beneficial ideas starts with individual creativity and ends with idea fruition (Perry-Smith & Mannucci, 2017). Although societal and contextual factors play a role (Shalley et al., 2004), each creative thought begins in the mind of the individual. Cognitive science has already highlighted that the ability to generate new innovative concepts may be strongly influenced by a human's knowledge base (Agogué et al., 2013). It seems that the creation of ideas that will not refer to familiar situations seems to be mostly challenging. It is directly related to a cognitive effect called "fixation effect" (Jansson and Smith, 1991). In order to solve such an effect, new knowledge should be introduced. Nevertheless, at this point, the nature of knowledge plays a crucial role. Depending on the kind of knowledge, new inputs can either promote fixation or foster generativity. As it was demonstrated, the limited knowledge associated with non-original conceptual pathways reduces generativity, whereas original knowledge inputs contribute to the improvement of the diversity and uniqueness of the generated ideas (Agogué et al., 2013). Further research study on the literature can lead us to the assumption that when people have complex knowledge structures – that is, a wide range of schemas rich in knowledge qualities and interschema linkages – and flexible knowledge structures – that is, weak linkages that allow for the inclusion of new domains and schemas – idea production is more successful.

The idea generation is more effective when individuals' knowledge structures are both complex (i.e., including a large set of schemas rich in knowledge attributes and inter-schema linkages) and flexible (i.e., including weak linkages and thus open to the addition of new domains and schemas). But then two questions arise:

When do breadth and depth of knowledge tend to foster creativity more or less?

Does the impact on creativity change as a person advances in their career?

1.3 KNOWLEDGE STRUCTURE AND ITS CHARACTERISTICS

The concept of "knowledge structure" originated in the 1970s within the field of "cognitive science," a field that developed from the intersection of computer science and psychology. This term had a crucial part in exploring the intricacies of the human mind. It was envisioned as a boundary concept capable of encompassing both the structures run by intelligent machines and human knowledge (Ley, 2020).

Three categories of human knowledge are known: (1) domain knowledge, (2) task knowledge, and (3) phenomenological knowledge. Ontological and specialized domain knowledge is usually taught in schools as topic matter. As in a thesaurus, domain-specific knowledge is made up of concepts and how they relate to one another. Task-dependent knowledge is acquired by using domain expertise to solve problems in scenarios as it is epistemological, strategic, and procedural. Procedural knowledge, for instance, distinguishes professionals from novices. Experience-based phenomenological knowledge comprises three types of knowledge that are only acquired through life: personal episodic knowledge, which is shared among a group of people, and tacit knowledge, which is automatic and not easily accessible to consciousness (Jonassen, 2000).

Knowledge structures represent how knowledge is organized within a domain; in many cases, they are also referred to as mental models, schemata, and cognitive maps (Goldsmith & Kraiger, 2013). They are achieved by determining the connections between concepts, ideas, and rules within a domain. To use knowledge efficiently and better absorb new information, people need to actively participate in organizing and representing what they learn. According to Ausubel et al. (1978), "What the learner already knows is the most important factor in learning." For new information to make meaning, it must be integrated into the existing knowledge structure. Hierarchical organization is a common structural representation of knowledge. It can be represented as a graph that goes from more general concepts to those that are more particular (a top-down configuration), resembling a reversed tree or organizational chart. Some representations, such as a temporal order or an interconnected network, have more complex structural designs. The diverse components of a particular domain knowledge may be portrayed using distinct structures, and an individual's memory may have a multitude of coexisting representations (Jonassen & Grabowski, 1993).

In education, knowledge structures are characterized as an understanding of the interconnections among concepts within a specific domain (Jonassen & Wang, 1993). These structures are utilized by intelligent educational systems for various purposes, such as adjusting navigation and facilitating activities like knowledge tracing, adaptive assessment, information presentation, and learner guidance within intelligent tutoring systems. Moreover, in conceptualizations of workplace learning, knowledge structures play a crucial role by representing distinct elements of the organizational knowledge base's structure.

These encompass not only connections between organizational concepts (like technology, marketing, and production), but also objectives, convictions, or a "cognitive framework that individuals apply to an information environment to structure and interpret it" (Ahuja & Novelli, 2015, p. 552).

1.4 CREATIVITY OF WORKERS AND THEIR KNOWLEDGE CHARACTERISTICS

The idea of creativity changed from a singular human condition to an outcome as studies on creativity developed. According to Gardner (1993), who proposed a modern widely recognized definition, a creative person is a person who consistently finds fresh solutions to issues, crafts products, or formulates new inquiries in a field, only for such solutions to eventually be accepted in a specific cultural context.

Since there is no one widely recognized definition or method for measuring creativity, researchers follow a multidimensional approach to the study of creativity (Caroff & Lubart, 2012; Agnoli et al., 2018; Wong et al., 2021). However, novelty and usefulness are two characteristics that are typically included in definitions of creativity (e.g., Woodman et al., 1993). Innovative concepts are special and valuable if they can help the company (Shalley et al., 2004). An idea must be useful to be creative. Although strange ideas can be unique, they may also be immoral or extremely difficult to implement in an organization (Shalley & Perry-Smith, 2001).

There are two knowledge attributes that have been proposed to encourage creativity. People should, on the one hand, possess depth of knowledge (i.e., the extent to which a person is informed about a particular domain). The complexity of knowledge structures intensifies with an increase in knowledge depth, offering individuals more schemas that are enriched with attributes and linkages. This results in a more expansive ideational sample within a specific domain (Mannucci & Yong, 2018). On the contrary, individuals also need knowledge breadth, which is perceived as the extent to which a person is knowledgeable in several different domains (Sosa, 2011). However, its overload can lead to creativity impairment (Wadhwa & Kotha, 2006). Knowledge breadth enhances the flexibility of knowledge structures by exposing individuals to diverse domains, thereby loosening existing connections and facilitating the formation of new ones within and across domains. As a result, both knowledge depth and knowledge breadth play a role in influencing creativity by shaping a person's mental cognitive structures (Mannucci & Yong, 2018).

As one gets older in their career, inter-domain links become stronger and knowledge structures become more rigid. The ability of a person to successfully integrate various knowledge domains to find unique combinations is influenced by their career age. Knowledge depth is less helpful for creativity later in life when people's knowledge structures become more inflexible. It is more helpful for creativity early in a person's career, when knowledge structures are relatively flexible and complexity needs to be increased. In later stages, when a strong stiffness and flexibility needs to be improved by loosening up knowledge structures, knowledge breadth is the most advantageous (Mannucci & Yong, 2018).

In the literature, it has also been investigated that a number of individual differences, such as demographic and biographical factors, have an impact on creativity. However, two characteristics, namely people's personalities and cognitive styles, have received the greatest attention (Johnson, 2003; Wu & Wu, 2020; Naseer et al., 2021). Openness to experience is the characteristic that most frequently correlates with creativity. People who are open tend to be unconventional, inquisitive, and broad-minded. The qualities most conceptually related to creative performance are openness, which is defined as being inquisitive, adaptable, inventive, and receptive to new concepts, experiences, and unusual viewpoints (Costa & McCrae, 1992). On the other hand, a man with poor openness tends to be traditional, uncreative, and unanalytical. Furthermore, open people are more likely to seek out novel circumstances that provide them with increased access to fresh experiences and viewpoints, as well as to be more adaptable in their ability to process and integrate new and unrelated information. These people also feel more compelled to look for novel circumstances that provide them more access to fresh viewpoints and experiences (Lebuda et al., 2021).

Across a range of fields, creativity is often positively correlated with openness to new experiences. For example, in the Five Factor Model of personality (Patterson et al., 2009), it is discovered that being open to new experiences and related concepts, i.e., curiosity, imagination, and wide interests, positively influences and may even be the most significant predictor of an individual's creative tendencies.

Numerous studies have looked into the relationship between a person's cognitive style and their creative outcomes. An individual's ideal cognitive style can encourage people to concentrate on crucial information and use their imagination to generate originality and creativity (Lomberg et al., 2017; Chen et al., 2018). The findings imply that people who exhibit an inventive style tend to be more imaginative than those who do it in an adapted way. Contextual variables are taken into account, including the complex nature of the work, relationships with managers and coworkers, rewards, evaluation, deadlines and goals, and the physical layout of the workspace. However, it has to be underlined that there are two factors that influence the relationship between employee creativity and extrinsic rewards (payment and recognition): the complexity of the employee's work and their cognitive style. For workers with an adaptable cognitive style, who perform relatively basic tasks, there is a favorable relationship between extrinsic rewards and creativity. However, there is a modest correlation between rewards and creativity for workers with an inventive cognitive style and who handle complex tasks. Moreover, for those in the adaptive style/complex job and innovative style/simple job situations, there is even a negative relationship (Baer et al., 2003).

Furthermore, a job's design has a significant role in fostering employee creativity. People, who work on complex tasks, are more probable to have high levels of intrinsic motivation, which prompts them to come up with innovative ideas. In particular, challenging occupations ought to increase people's enthusiasm for and drive to complete their work, and this enthusiasm ought to stimulate creativity (Falk, 2023).

According to Yoo et al. (2019), it is possible to notice the connection between three aspects of a job: skill diversity, autonomy, and feedback; three aspects of the corporate context: organizational climate, resources, and extrinsic rewards; and individual creativity. Every aspect of a profession has a favorable impact on an individual's creativity. The relationship between job characteristics and individual creativity was only significantly moderated by extrinsic rewards; these rewards had a favorable impact on the correlation between individual creativity and autonomy, but an unfavorable impact on the correlation between individual creativity and skill variety.

The relationship between leadership style and staff creativity has been the subject of numerous studies. According to intrinsic motivation theory, encouraging leadership style should increase intrinsic motivation. Employees who work with loving and helpful coworkers are also predicted to display high levels of creativity since it increases intrinsic motivation (Al Harbi et al., 2019). Through the subordinates' creative role identity, there was a positive indirect association between the creativity of supervisors and that of their subordinates; this relationship was stronger when employees felt that there was more organizational support for creativity (Koseoglu et al., 2017). Furthermore, organizational innovation and followers' creativity are significantly positively correlated with transformational leadership. Additionally, a strong positive correlation between both of them is discovered. Furthermore, it is noticed that there is a positive and substantial association through the mediating functions of psychological empowerment, support for innovation, workplace connections, and employee learning between followers' creativity and transformational leadership. The information did, however, indicate that the relationship between creativity and transformative leadership is not much impacted by intrinsic drive (Al Harbi et al., 2019).

Work activities would be neglected in favor of evaluation, which would reduce intrinsic drive and, in turn, innovation. It implies that people should view developmental assessments as informative and supportive, promoting active involvement which will lead to increased creativity (Hickeman, 2023). People are supposed to feel under pressure to reach production targets or deadlines when they are present, which lowers their intrinsic motivation and inventiveness. Many unwelcome or unanticipated interpersonal intrusions were encountered by people who worked in crowded environments with little boundaries, which subsequently had an impact on their attitudes and behaviors. It is clear that creativity takes place in a physical context. Such a place both limits and permits the unrestricted flow of sensory experiences and interpersonal proximity. Certain sensory experiences, such as viewing the source material, sight, and sound (including noise), may become available due to the confines. This framing permits some cognitive processes while limiting others. It could elicit feelings that either increase or decrease the ability to be creative (Kristensen, 2004).

1.5 PERSONAL AND CONTEXTUAL CHARACTERISTICS OF SMART OPERATOR

According to Deloitte, in 2018, humans carried out 71% of tasks, but that percentage will drop in the near future which will result in employment losses. However, the research also states that 133 million new employment will be generated in addition to the 75 million jobs that would be removed. A lot of these new 133 million jobs will need not only proficiency with the newest digital technologies but also a new set of soft skills like creativity and critical thinking (ERP Today, 2023). Most of them will be performed using smart manufacturing systems that use cloud computing infrastructures and the internet of things to create an intelligent network where the physical environment is deeply entwined with the matching cyber twin. From this angle, operators serve as a bridge to enable the best possible integration of virtual and physical assets. Because of this, human factors are at the center of a positive feedback loop that is closed-loop and helps the production system as a whole develop and change over time (Ciccarelli et al., 2022). Therefore, they need such skills and capabilities as being able to define, design, deploy, and refine conceptual and physical solutions aimed at generating value for Industry, which clearly refers to the levels of knowledge depth. On the other hand, it is required that smart operator has a knowledge in the domains of operations research and operations management, maintenance, quality management systems, modeling and simulation, mechatronics and automation, robotics and artificial intelligence, project management, information systems, systems engineering, robotics and modeling, logistics and supply chain management, which refers to knowledge breadth (Dash et al., 2019).

The complexity of smart operators' everyday work increases, thus they must be extremely adaptable and show that they can adapt while using their depth and breadth of knowledge in a very dynamic working environment. They need to use critical and creative thinking, be collaborative, take initiative, be able to use social-emotional learning, work in a team, be self-confident, aware of culture, innovative, make decisions, have good communication skills, and change their mindset for lifelong learning. They need to use logical and mathematical reasoning, visualization and show sensitivity to problems. As their tasks are complex, they have to possess general and interdisciplinary understanding of technology, specific knowledge of manufacturing operations and procedures, and machine technical know-how to perform maintenance-related tasks. Moreover, the greatest value and driving force for industrial engineers is the welfare of people; additional values include sustainability, efficiency, justice, and well-being of people. In order to cope with deadlines and goals, time management is required.

The operators who most actively promote their ideas are also far more likely to be the ones displaying the strongest idea generation activities. Although an unstructured team innovation setting benefits more from having team members display such extraordinary behaviors, organized idea journey arrangements produce superior team level inventive solutions when idea championing behaviors are more evenly distributed across team members. Teams with members who are excellent in both idea generation and championing are able to transform creative ideas into innovative team solutions. As a result, it is crucial to first educate managers on the value of idea championing. When putting teams together, they should consider integrating people who can generate ideas as well as those who can market, "sell," and build coalitions to support them through implementation. Aside from choosing these people, businesses could also train people in creativity improvement techniques, creative thinking, and social skills – the ability to influence others – to help them generate and push ideas more effectively (Černe et al., 2024).

However, it may happen that teams could lack members who are particularly good at coming up with and promoting ideas. But managing the ensuing efforts to promote and implement ideas is simpler than managing the idea generating process itself (Černe et al., 2024). It is also important to highlight that managers should pay attention to organizing the innovation process because hierarchical organizations typically hinder creativity (Kim & Zhong, 2017). Hence, even in cases where teams' initial levels of idea production and championing are lower or more distributed, an organized idea journey founded on an upgrade of thinking modes appears to provide higher levels of team innovation output (Černe et al., 2024).

1.6 CONCLUSIONS

Few contextual variables were empirically investigated because the majority of the examined research were theoretical or small-scale case studies. However, it led us to the final remarks that skills and technology acquisition are important personal and contextual factors that can increase creativity in relevance to the implementation of Industry 4.0. The operators who are able to promote their ideas enthusiastically are also far more likely to be the ones displaying the strongest idea generation activities.

REFERENCES

Agnoli, S., Vanucci, M., Pelagatti, C., Corazza, G.E. (2018). Exploring the link between mind wandering, mindfulness, and creativity: A multidimensional approach. *Creativity Research Journal*, 30(1), 41–53.

Agogué, M., Kazakçi, A., Hatchuel, A., Le Masson, P., Weil, B. (2013). The impact of type of examples on originality: Explaining fixation and stimulation effects. *The Journal of Creative Behavior*, 48(1), 1–12.

Ahmed, J., Mrugalska, B., Akkaya, B. (2022). Agile management and VUCA 2.0 (VUCA-RR) during Industry 4.0. In: Akkaya, B., Guah, M.W., Jermsittiparsert, K., Bulinska-Stangrecka, H., Kaya, Y. (eds.) *Agile Management and VUCA-RR: Opportunities and Threats in Industry 4.0 towards Society 5.0* (pp. 13–26). Leeds: Emerald Publishing Limited.

Ahuja, G., Novelli, E. (2015). Knowledge structures and innovation: Useful abstractions and unanswered questions. In: Easterby-Smith, M., Lyles, M.A. (eds.) *Handbook of Organizational Learning and Knowledge Management* (pp. 551–578). Hoboken, NJ: John Wiley & Sons.

Al Harbi, J.A., Alarifi, S., Mosbah, A. (2019). Transformation leadership and creativity: Effects of employees pyschological empowerment and intrinsic motivation. *Personnel Review*, 48(5), 1082–1099.

Ausubel, D.P., Novak, J.D., Hanesian, H. (1978). *Educational Psychology: A Cognitive View*. New York: Holt, Rinehart and Winston.

Baer, M., Oldham, G.R., Cummings, A. (2003). Rewarding creativity: When does it really matter? *The Leadership Quarterly*, 14(4–5), 569–586.

Caroff, X., Lubart, T. (2012). Multidimensional approach to detecting creative potential in managers. *Creativity Research Journal*, 24(1), 13–20.

Černe, M., Kaše, R., Škerlavaj, M. (2024). Idea championing as a missing link between idea generation and team innovation implementation: A situated emergence approach. *European Management Journal*, 42(2), 233–244.

Chen, M.H., Chang, Y.Y., Lin, Y.C. (2018). Exploring creative entrepreneurs' happiness: Cognitive style, guanxi and creativity. *International Entrepreneurship and Management Journal*, 14, 1089–1110.

Ciccarelli, M., Papetti, A., Cappelletti, F. (2022). Combining world class manufacturing system and Industry 4.0 technologies to design ergonomic manufacturing equipment. *International Journal on Interactive Design and Manufacturing*, 16, 263–279.

Costa, P.T., McCrae, R.R. (1992). *Revised NEO Personality Inventory (NEO PI-R) and NEO Five-Factor Inventory (NEO-FFI) Professional Manual*. Odessa, FL: Psychological Assessment Resources.

Dash, D., Farooq, R., Panda, J.S., Sandhyavani, K.V. (2019). Internet of Things (IoT): The new paradigm of HRM and skill development in the fourth industrial revolution (Industry 4.0). *IUP Journal of Information Technology*, 15(4), 7–30.

Ejsmont, K. (2021). The impact of Industry 4.0 on employees-insights from Australia. *Sustainability*, 13(6), 3095.

ERP Today (2023). *Industry 4.0: A people - Centric Transformation*. https://erp.today/industry-4-0-a-people-centric-transformation. Retrieved on November 26, 2023.

Falk, S. (2023). *Understanding the Power of Intrinsic Motivation*. Harvard Business Review. https://hbr.org/2023/03/understand-the-power-of-intrinsic-motivation. Retrieved on December 08, 2023.

Gardner, H. (1993). *Creating Minds: An Anatomy of Creativity Seen through the Lives of Freud, Einstein, Picasso, Stravinsky, Eliot, Graham, and Gandhi*. New York: Basic Books.

Goldsmith, T., Kraiger, K. (2013). Applications of structural knowledge assessment to training evaluation. In: Ford, J.K. (eds.) *Improving Training Effectiveness in Work Organizations* (pp. 73–96). New York: Routlegde.

Hickeman, R. (2023). Assessment, creativity and learning: A personal perspective. *Future in Educational Research*, 1(2), 104–114.

Jansson, D.G., Smith, S.M. (1991). Design fixation. *Design Studies*, 12(1), 3–11.

Johnson, R.C. (2003). *Study of the Relationship between Cognitive Styles of Creativity and Personality Types of Military Leaders*. Oklahoma: The University of Oklahoma.

Jonassen, D. (2000). Knowledge is complex: Accommodating human ways of knowing. In: Soergel, D., Srinivasan, P., Kwasnik, B. (eds.) *Proceedings of the 11th ASIS&T SIGKR Classification Research Workshop* (pp. 1–7). Silver Spring, MD: American Society for Information Science.

Jonassen, D.H., Grabowski, B.L. (1993). *Handbook of Individual Differences: Learning and Instruction*. Hillsdale, NJ: Lawrence Earbaum Associates.

Jonassen, D.H., Wang, S. (1993). The physics tutor: Integrating hypertext and expert systems. *Journal of Educational Technology Systems*, 22(1), 19–28.

Kim, Y.J., Zhong, C.-B. (2017). Ideas rise from chaos: Information structure and creativity. *Organizational Behavior and Human Decision Processes*, 138, 15–27.

Koseoglu, G., Liu, Y., Shalley, C.E. (2017). Working with creative leaders: Exploring the relationship between supervisors' and subordinates' creativity. *The Leadership Quarterly*, 28(6), 798–811.

Kristensen, T. (2004). The physical context of creativity. *Creativity and Innovation Management*, 13(2), 89–96.

Lebuda, I., Karwowski, M., Galang, A.J.R., Szumski, G., Firkowska-Mankiewicz, A. (2021). Personality predictors of creative achievement and lawbreaking behavior. *Current Psychology*, 40, 3629–3638.

Ley, T. (2020). Knowledge structures for integrating working and learning: A reflection on a decade of learning technology research for workplace learning. *British Journal of Educational Technology*, 51(2), 331–346.

Lomberg, C., Kollmann, T., Stöckmann, C. (2017). Different styles for different needs-The effect of cognitive styles on idea generation. *Creativity and Innovation Management*, 26(1), 49–59.

Mannucci, P.V., Yong, K. (2018). The differential impact of knowledge depth and knowledge breadth on creativity over individual careers. *Academy of Management Journal (AMJ)*, 61(5), 1741–1763.

Mrugalska, B., Ahmed, J. (2021). Organizational agility in Industry 4.0: A systematic literature review. *Sustainability*, 13, 8272. https://doi.org/10.3390/su13158272.

Mrugalska, B., Stasiuk-Piekarska, A. (2020). Readiness and maturity of manufacturing enterprises for Industry 4.0. In: Mrugalska, B., Trzcielinski, S., Karwowski, W., Di Nicolantonio, M., Rossi, E. (eds.) *Advances in Manufacturing, Production Management and Process Control. AHFE 2020.* Advances in Intelligent Systems and Computing, vol. 1216 (pp. 263–270). Cham: Springer.

Naseer, S., Khawaja, K.F., Qazi, S., Syed, F., Shamim, F. (2021). How and when information proactiveness leads to operational firm performance in the banking sector of Pakistan? The roles of open innovation, creative cognitive style, and climate for innovation. *International Journal of Information Management*, 56, 102260.

Patterson, F., Kerrin, M., Gatto-Roissard, G. (2009). *Characteristics & Behaviours of Innovative People in Organisations* (pp. 1–63). London: NESTA Policy and Research Unit (NPRU).

Perry-Smith, J.E., Mannucci, P.V. (2017). From creativity to innovation: The social network drivers of the four phases of the idea journey. *Academy of Management Review*, 42(1), 53–79.

Piwowar-Sulej, K. (2020). Human resource management in the context of Industry 4.0. *Organization & Management Scientific Quarterly*, 1, 103–113.

Shalley, C.E., Perry-Smith, J.E. (2001). Effects of social-psychological factors on creative performance: The role of informational and controlling expected evaluation and modeling experience. *Organizational Behavior and Human Decision Processes*, 84(1), 1–22.

Shalley, C.E., Zhou, J., Oldham, G.R. (2004). The effects of personal and contextual characteristics on creativity: Where should we go from here? *Journal of Management*, 30, 933–958.

Sollosy, M., McInerney, M., Braun, C.K. (2016). Human capital: A strategic asset whose time has come to be recognized on organizations' financial statements. *Journal of Corporate Accounting & Finance*, 27, 19–27.

Sosa, M.E. (2011). Where do creative interactions come from? The role of tie content and social networks. *Organization Science*, 22, 1–21.

Umair, S., Waqas, U., Al Shamsi, I.R., Kamran, H., Mrugalska, B. (2023a). Impact of strategic orientation and supply chain integration on firm's innovation performance: A mediation analysis. In: Mrugalska, B., Ahram, T., Karwowski, W. (eds.) *Human Factors in Engineering* (pp. 85–103). Boca Raton, FL: CRC Press.

Umair, S., Waqas, U., Mrugalska, B., Al Shamsi, I.R. (2023b). Environmental corporate social responsibility, green talent management, and organization's sustainable performance in the banking sector of Oman: The role of innovative work behavior and green performance. *Sustainability*, 15, 14303.

Umair, S., Waqas, U., Mrugalska, B. (2023c). Cultivating sustainable environmental performance: the role of green talent management, transformational leadership, and employee engagement with green initiatives. *Work: A Journal of Prevention, Assessment & Rehabilitation,* 1–13.

Wadhwa, A., Kotha, S. (2006). Knowledge creation through external venturing: Evidence from the telecommunications equipment manufacturing industry. *Academy of Management Journal*, 49, 819–835.

Wong, C.C., Kumpulainen, K., Kajamaa, A. (2021). Collaborative creativity among education professionals in a co-design workshop: A multidimensional analysis. *Thinking Skills and Creativity,* 42, 100971.

Woodman, R.W., Sawyer, J.E., Griffin, R. (1993). Toward a theory of organizational creativity. *Academy of Management Review*, 18, 293–321.

Wu, T.T., Wu, Y.T. (2020). Applying project-based learning and SCAMPER teaching strategies in engineering education to explore the influence of creativity on cognition, personal motivation, and personality traits. *Thinking Skills and Creativity*, 35, 100631.

Yoo, S., Jang, S., Ho, Y., Seo, J., Yoo, M.H. (2019). Fostering workplace creativity: Examining the roles of job design and organizational context. *Asia Pacific Journal of Human Resources*, 57(2), 127–149.

Yousefi, S., Derakhshan, F., Karimipour, H. (2020). Applications of big data analytics and machine learning in the Internet of Things. In: Choo, K.K., Dehghantanha, A. (eds.) *Handbook of Big Data Privacy* (pp. 77–108). Cham: Springer.

2 Navigating Skills Shortages
Bridging Human Factors and Fashion Practice Research for Collaborative Innovation in UK Fashion Manufacturing

Kat Thiel and Susan Postlethwaite

2.1 INTRODUCTION

This paper builds on the conference paper "Human-centric research of skills and decision-making capacity in fashion garment manufacturing to support robotic design tool development" (Thiel and Postlethwaite, 2023), first presented at the 14th International Conference on Applied Human Factors and Ergonomics. The previous paper was divided into two parts, whereas this paper provides an extension to only part two, providing an understanding of how expertise in sewing machinists is captured and used to assess implications for tooling design for advanced fashion manufacturing. Part one of the conference paper presented a literature review of the existing skills within the technical workforce in UK fashion garment manufacturing, detailing the challenges imposed on the fashion sector by ongoing skills shortages and the resulting difficulties in recruiting talent due to low incentives and inadequate training options. Data was drawn from five key reports, which underscored the need for innovative manufacturing processes and advanced training while stressing the need for action against the backdrop of a rapidly declining skilled workforce (The Alliance Project Team, 2015; Postlethwaite et al., 2022; Hooper et al., 2022; West Yorkshire Combined Authority, 2021; The Environmental Audit Committee, 2019; Harris et al., 2021). Ongoing Government support for advanced manufacturing in the UK suggests that it would be possible to form pathways that could include fashion manufacturing, and by extension, creating an opportunity for fashion education to evolve by incorporating Industry 4.0 principles to increase flexibility, collaboration, and resilience when training new talent. A transition from STEM to STEAM+D, as proposed by the Design Council's report "Designing a Future Economy – Developing Skills for Productivity and Innovation" (2018), advocates for a productive fusion of practical ability and creativity. Incentives for universities to offer multidisciplinary

DOI: 10.1201/9781003505327-2

design courses are recommended to engage with the fourth and fifth industrial revolution, combining science, engineering, and Human Factors with design degrees to boost graduate's skillsets. To explore Industry 4.0 and 5.0, a future workforce will require a blend of hard and soft skills, including technological understanding, critical thinking, and interdisciplinary collaboration. The paper further argued that effective investment in R&D is essential for crafting agile tools that align with the specific needs of small to medium enterprises. If combined with robotics, a people-centred approach to automation, focusing on Human Factors and Fashion Practice Research, is therefore crucial for the development of this emerging field.

The following is a detailed discussion of a collaboration between researchers from Manchester Fashion Institute and researchers from the Psychology and Human Factors Group at Cranfield University. As an exploratory study developing a new research approach, the project is based on Human Factors and Fashion Practice Research. The latter is an emerging field of research that is distinct from practice-based fashion which interrogates fashion practice from a humanities perspective deploying a canon of practice-based methods and reflections on form-giving processes, often from an individual's perspective. While theoretical frameworks for wider design practice research exist (Frayling, 1993; Gaver, 2012; Vear, 2021), they often deliberately exclude engineering and social sciences (Koskinen et al., 2011). Laurene Vaughan's concept of the designer-practitioner-researcher (2019) establishes a new understanding of some of these paradoxical positions; this term describes hybrid practitioners who focus their work on technological inquiry aimed at transforming existing systems in need of re-evaluation through design. Building on this and encouraging a transdisciplinary approach to research, we are proposing a Fashion Practice Research that supports the design and study of practice enabling technology based in Design Anthropology, harnessing the emic insight fashion designer-practitioner-researchers offer to modernise the very systems they operate in and consequently challenge fashion's Programmatic Tradition (Krogh and Koskinen, 2022). Design Anthropology as an emergent field understands designing and using as inseparable practices that directly inform skilled work (Gunn and Donovan, 2016), hence providing a fitting framework for collaborative, future-making research with social science researchers. Designing flexible systems that allow multiple users to find their optimum way of operating a machine then becomes a key consideration in the development stages of new technology for agile manufacturing. Studying decision-making activity during sewing tasks is the first step in establishing a nuanced understanding of how sewing machinists deploy their various skills and knowledges, as well as how these can be translated into new technology.

This paper presents a detailed insight into the first collaborative endeavours aimed at understanding how Human Factors methods may inform the design of appropriate tolling for low-volume high-value manufacturing, thereby supporting the emergence of a place-based fashion industry in which designer manufacturers work alongside cobotic technologies. The paper details and reflects on the outcomes of physiological eye-tracking tests conducted with a limited sample group, examining the cognitive decision-making processes observed during a conventional sewing assembly task. It subsequently discusses the findings, limitations, and implications of the study (Figure 2.1).

Sewing machinists

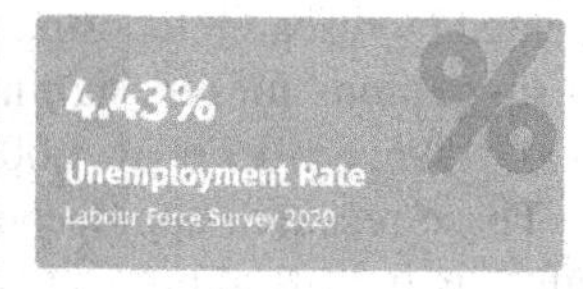

Future employment projections (Working Futures UK)

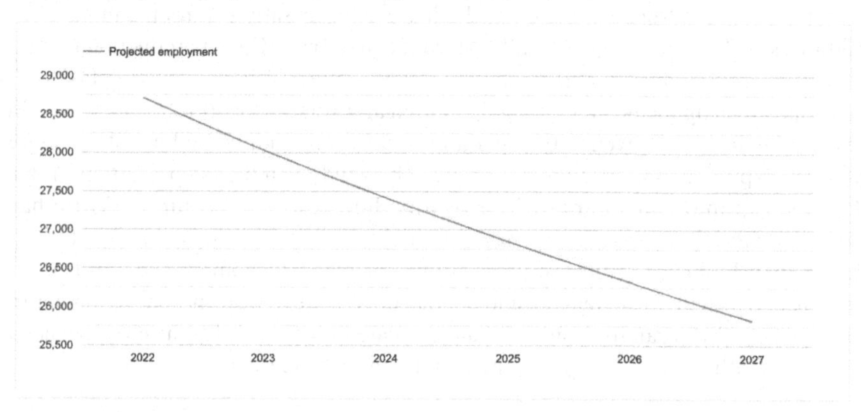

FIGURE 2.1 Union Learn – Future employment projections sewing machinist (accessed January 2023).

2.2 UNDERSTANDING EXPERTISE IN SEWING MACHINISTS AND ITS IMPLICATIONS FOR TOOLING DESIGN IN FASHION MANUFACTURING

2.2.1 Background

The aforementioned reports suggest measures to increase skills in young people by making relevant training available more broadly. However, very few papers concentrate on evidencing skills levels in sewing machinists. Recent research by Küçük and Birol (2021) developed a competency analysis system (CAS) assessing each sewing machinist of an assembly plant against a set of six criteria. The scale they developed encompassed not only the skill or competence of the machinist (expressed in the criteria Machine Knowledge, Seniority Level, and Performance), but also human resources factors such as Senior View, Absenteeism, and Overtime Availability. The aim of the CAS approach is the improvement of the overall speed and accuracy of an assembly line known as line balancing. The researchers found that senior, more qualified staff generally received higher scores, whereas newer staff were ranking lower. Recommendations were made to implement more training activities to level the operator scores across assembly lines to avoid bottlenecks in the production. Globally operating, large-scale garment production has captured researchers' attention for many decades, predominantly addressing poor posture and working conditions. Ongoing research has been dedicated to factors causing musculoskeletal disorders (Blader et al., 1991; Brohi et al., 2022; Das and Natarajan, 2022), complemented by research focusing on ergonomic factors during sewing activities resulting in suggestions for redesigning the sewing workplace (Delleman and Dul, 1989; Li' et al., 1995; Sarder et al., 2007; Chan et al., 2002; Ahmad et al., 2021).

These research approaches generally aim to increase productivity and operator safety in industrial production sites offshore where operators are commonly responsible for one task in an assembly line. This is largely not representative of the fashion manufacturing landscape in the UK, which is predominantly still rooted in micro- to medium-sized companies and small Cut, Make and Trim (CMT) production units. A sewing machinist in the UK is likely in charge of many, if not all, sewing tasks of a garment, stressing the need for a highly skilled and versatile workforce. However, little attention has been given to the operational skills and individual know-how aggregated by expert sewing machinists to date. This study is a first step at attempting to close the gap in the literature by investigating how sewing machinists' skilful handling of material and machines can be studied to inform the design of new tooling. Rather than looking to improve productivity through ergonomics, the remit of this work is concerned with ways to understand how more rewarding human–machine interactions (HMI) can be designed that stimulate the operator by enriching an activity that can otherwise be very repetitive and straining on the body.

2.2.2 Method

This study follows a Method Stacking approach to data gathering (Thiel and Eimontaite, 2023). In Method Stacking, a transdisciplinary team of researchers is formed and complemented by skilled practitioners who become active agents in the research. Method Stacking aims to iteratively stack expertise as well as objective and subjective methods, always starting with the physiological data capture in loosely controlled, real-life production environments. Reflective interviews are held with the participants after data is analysed. As a consequence of loose control, fluctuations and irregularities in the data sets are more likely to occur, highlighting the limitations of any method. In Method stacking, methods are not seen as always returning truthful data, nor are they discounted when deviations occur. What would be considered corrupted data in most studies becomes the foundation of rich qualitative discussions with the expert makers, stacking evidence gradually as the interrogation deepens. The physiological data is probed and reviewed, and knowledge is co-constructed by the researchers together with the expert makers. Working with limitations and deviations constructively demands a high level of responsiveness as both the evidence and methods are being challenged and cross-verified.

2.2.2.1 Participants

Participants in the trial sample were recruited on a voluntary basis. The volunteers were expertly trained sewing machinists who worked in either academic or technical roles within fashion departments at the Royal College of Art in London and Manchester Metropolitan University respectively. Initially, four participants were recruited, with one test failing to return usable data and the participant subsequently being excluded from the sample. This left three participants' data sets. Those three participants were female, spectacle wearers, and between 30 and 45 years old.

2.2.2.2 Procedure

Prior to testing, participants signed a consent form and were briefed on the procedure of the physiological data capture. They were equipped with an eye-tracking device by SensoMotoric Instruments (SMI) and an Empatica E4 wristband to capture heart rate and electrodermal activity during task performance. Task performance and data capture started after a quick calibration process. The overall goal was the fabrication of a standard sleeve placket on an industrial sewing machine. Participants were allowed to choose their preferred machine freely. As all of them were spectacle wearers and the SMI device did not include additional prescription lenses, the team decided on a mixed approach, with participants either leaving their prescription glasses on or taking them off if they felt they could confidently perform the task without them. Since the eye-tracking device is wired directly to a laptop when in use, a restraint was imposed to reduce movement during the task performance. Traditionally, placket construction involves pressing the placket halfway through the task. However, the machinists were instructed to remain at their sewing machine for the whole duration of the task meaning they could not be using a pressing iron in between sub-tasks.

The tests were carried out with no specific hypothesis guiding the trials. Rather, the experiments served as a vehicle to understand the value of specific quantitative methods to evidence skill in a creative environment that could not be fully controlled. Working within changing environments and with a range of materials logically challenges the collection of data. Crucially, being accustomed to variabilities is the nature of design work and research. Multiple possibilities occur frequently, and shifting attitudes towards materials, methods, and techniques are the norm (Vaughan, 2019; Krogh and Koskinen, 2022). After the eye-tracking, data was coded for Areas of Interest (AOI) and a Hierarchical Task Analysis (HTA) established, and the sewing machinists were reengaged for reflective interviews. Participants were presented with playbacks of their performance as captured by the eye-tracking device to collect the motivations for decisions made during task performance. The interviews shed further light on the participants' overall perception of the eye-tracking accuracy, which became a crucial element in evaluating how deviations and inaccuracies in the quantitative data sets spark more useful qualitative insight for the subject of the study. The Empatica E4 device for heart rate and sympathetic nervous system activity functioned perfectly, but it ceased to return data relevant to the study of skill and will therefore be neglected in the reporting of this paper.

2.2.3 Analysis

2.2.3.1 Areas of Interest

The analysis of the eye-tracking data across all three trials indicates two strong AOIs. Figure 2.2 presents a breakdown of participant 3's three trials, identifying the fabric (dotted lines) as the prominent focal point during all three tests. The second largest AOI, horizontal lines, is the area around the presser foot where the needle moves in and out of the fabric. Activities coded in horizontal lines therefore indicate the time spent sewing.

Dotted "fabric" areas indicate activities related to visual inspection and manual manipulations of the fabric in preparation for the next seam. These handling tasks can be a combination of any of the following: unpicking a seam, pinning

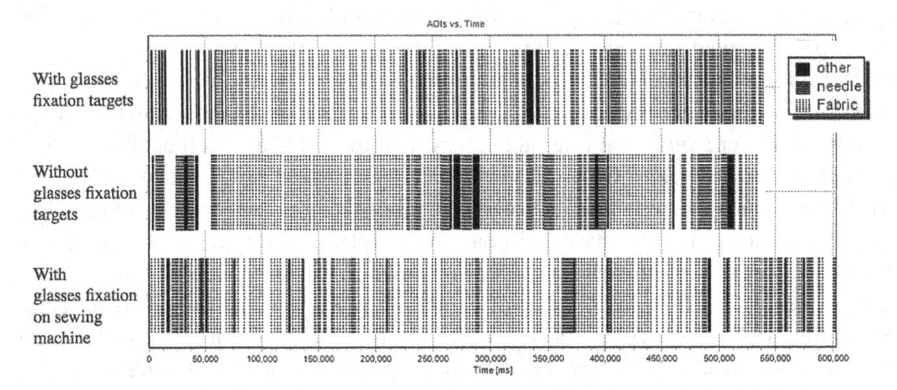

FIGURE 2.2 AOIs over time, three plackets sewn by participant 3. With and without prescription glasses and varying fabrics.

and unpinning, trimming, cutting, folding, turning, aligning, marking, and measuring the fabric. All these sub-tasks are carried out directly on the sewing machine table to save time and ensure efficiency. By comparing these three tests, it is obvious that the focus tracker seems to lose connection, as indicated by unaccounted white interferences, especially in tests one and three. The same phenomena can be observed clearly in the video recording, where the focal tracker erratically flares to the upper left corner. This is attributed to the fact that participant 3 wore spectacles underneath the eye-tracker during these tasks. Test two, without prescription glasses, returned much more stable and consistent data, but the quality of the placket suffered. Combinations of lens flares and varying light conditions of both natural and artificial light on and around the sewing machines were observed during trials, which might have affected the consistent capture of eye-tracking data; these light sources are however an essential requirement in factories relying on human machinists. Another significant observation is the speed with which the three tasks were carried out. Tests 1 and 3 were both done with prescription lenses; however, test three took significantly longer to complete. The defining factor here is the change to a different material, from a light calico with high contrast in tests 1 and 2 to a dark blue colour in test 3 (see Table 2.1). While the AOIs give valuable insights to support an assessment of the perfect condition for eye-tracking tests during sewing activity, they show little evidence of skilful handling of the task beyond overall speed and dwell time.

2.2.3.2 Hierarchical Task Analysis

While the AOIs indicated sustained visual focus on the task and showed a dwelling rate in two main areas, the analysis further involved a Hierarchical Task Analysis to discern what types of activities were carried out more specifically by each of the sewing machinists. The overall goal of fabricating a standard sleeve placket can be broken down into five main steps:

TABLE 2.1
All Participants' Conditions Compared

	Machine	Fabric	Location	Spectacles under Eye-Tracking Device	Eye-Tracking Fixation Shifts	Cross Markers for Calibration
Participant 1	1	Pale Blue shirting	Royal College of Art	Yes	Consistently off focus	Yes
Participant 2	2	Pale Blue shirting	Royal College of Art	No	Occasionally off focus	Yes
Participant 3	3	Calico	Manchester Metropolitan University/MFI	Yes	Frequent erratic flares	Yes
Participant 3	3	Calico	Manchester Metropolitan University/MFI	No	Occasionally off focus	Yes
Participant 3	3	Dark blue shirting	Manchester Metropolitan University/MFI	Yes	Frequent erratic flares	No

1.0 Placing the placket
2.0 Securing the placket
3.0 Opening and turning the placket
4.0 Inner placket construction
5.0 Outer placket construction

During reflective interviews, participants described how the five hierarchical tasks of a standard placket were followed. Table 2.2 gives a detailed breakdown of participants 1's and 2's steps in direct comparison. They both shared sub-tasks, but added a significant number of sub-task variations. Not only did the sub-task variations differ in order and number, for example, participant 1 used one more step to complete the placket than participant 2, but they also introduced different techniques and tools to handle the same fabric with the same task. It is worth noting that according to the participants, decisions during the tests are the consequence of two guiding principles. They either try to find a way to make the task easier or are looking to increase the overall quality of the outcome. Which techniques and steps will be taken is assessed depending on three factors. Firstly, they assess how familiar they are with a machine; secondly, they judge the properties of the fabric; and lastly, they weigh material and machine in relation to task complexity.

2.2.4 Discussion

2.2.4.1 Cognitive Decision-Making

Sewing processes are combinations of complex 2D and 3D manipulations of textile material. Prospective sewing machinists go through training stages introducing them not only to increasingly more complex sewing tasks, but they also learn

TABLE 2.2
Direct Hierarchical Task Analysis Comparison between Participants 1 and 2, Including Shared Sub-Tasks and Individual Sub-Task Variations

	Shared Sub-Tasks	Participant 1 Sub-Task Variation	Participant 2 Sub-Task Variation
1.0 Placing the placket		1.1.1 No pre-pressed placket edges	1.1.2 Pre-pressed placket edges
	1.2 Placket fabric is pinned to sleeve and placed correctly		
2.0 Securing the placket	2.1 Back stitch at the start of the seam		
	2.2 Stitching the placket to the sleeve around the placement line, half a machine foot wide		
		2.3.1 Back stitch before and after each corner to give stability to the seam	
	2.4 Lifting the foot and turning the fabric to create the corners while needle remains in the fabric, creating a box shape		
	2.5 Back stitch at the end of the seam		
3.0 Opening and turning the placket			3.1.1 Removal of pins
	3.2 Straight cut with scissors on the placement line and cutting a triangle at the top towards the two corners		
		3.3.1 Removal of pins	

(*Continued*)

TABLE 2.2 (*Continued*)
Direct Hierarchical Task Analysis Comparison between Participants 1 and 2, Including Shared Sub-Tasks and Individual Sub-Task Variations

	Shared Sub-Tasks	Participant 1 Sub-Task Variation	Participant 2 Sub-Task Variation
	Restriction: NO PRESSING IRON		
4.0 Inner placket construction		4.1.1 Sewing a stay stitch at the narrow and wide edge of the placket instead of ironing	4.1.2 Using fingernails to crease the placket fabric along the stitch-lines instead of ironing this part
	4.2 Turning placket inside out and pushing seam allowance into the narrow placket		
		4.3.1 **No pins**, just folding placket along the stay stitch and using fingers to guide it to lay flat on the previous stitch-line	4.3.2 Pinning the narrower placket side down to the sleeve to cover the original stitch-line precisely, along the **pre-pressed placket edge**
	4.4 Placing folded narrow placket under machine at the top		
	4.5 Backstitch at the top, straight stitch-line down to the cuff end and backstitch to secure at the end		

(*Continued*)

TABLE 2.2 (*Continued*)

Direct Hierarchical Task Analysis Comparison between Participants 1 and 2, Including Shared Sub-Tasks and Individual Sub-Task Variations

	Shared Sub-Tasks	Participant 1 Sub-Task Variation	Participant 2 Sub-Task Variation
5.0 Outer placket construction		5.1.1 Turning triangle up and securing it to the placket with a stitch-line	
		5.2.1 Boxing out top of the outer placket, pushing corner out with scissors, and pressing flat with fingernails	
		5.3.1 **No pins just tension** Pushing all seam allowances into the outer placket and aligning the stay-stitch with the previous stitch-line	5.3.2 Pinning of pre-pressed outer placket edge to the previous stitch-line to hide all seam allowances
			5.4.1 Visual check on the back side of the fabric where to put pin for the box-turn
	5.5 Placing fabric under the machine, starting at the cuff end		
	5.6 Final seam: backstitch, straight line topstitching along the edge of the placket		
			5.7.1 Stopping before the pins to take them out, bar one that gets sewn over gently so that the placket stays in place
	5.8 Lifting the foot and turning the fabric to create the corners while needle remains in the fabric		
		5.9.1 Feeling with the tips of the fingers where the turning point is	5.9.2 Last pin indicates the last turning point to close the placket
	5.10 Backstitch to finish the seam		

how to judge the right machine calibration for each material they are working with. This includes tension detection and thread tension adjustments, choosing the correct needle according to materials, and choosing the right supporting tools for each task (e.g., scissors, pins, clips, and chalk marks). While most sewing techniques are broken down into individual steps that can be learned and followed, each machinist will develop a unique sensitivity to the materials they are working with. Over time and with experience, they learn how to combine their knowledge of materials, tools, and machines in their own terms, enabling them to produce high-quality finishes.

In the literature, this intimate knowledge based on experience is often described as tacit knowledge (Polanyi, 1958). Longstanding views that tacit knowledge, in contrast to explicit knowledge, is challenging or even impossible to formalise or pass on have been contested in recent years (Ray, 2009; Atkinson, 2022; Johnson et al., 2019). Smith (2001) suggests that its implicit, subjective, and preferential nature makes it akin to intuition. Atkinson, in particular, challenges the understanding that tacit knowledge in fashion designers and garment makers cannot be expressed and gives evidence of nonverbal or even digital communication styles that make it possible to pass on implicit, personal understandings of material and processes. In this way, tacit knowledge, and by extrapolation, tacit skill can be understood as a type of experiential knowledge/skill that can be captured and shared either through demonstration or other nonverbal acts. Experienced sewing machinists will respond to tactile feedback from both the machine and the fabric. With practice, they acquire expert hand-eye coordination and the ability to instinctively plan a procedure to ensure an expert execution of a task.

Interviews with participants revealed how different approaches and techniques were applied during the placket construction. Through detailed descriptions of their actions and intentions, all three participants demonstrated haptic skills (Smith, 2012), which are essential to accurately execute the task. This includes acute knowledge of the task steps, the speed and accuracy of the machine they are using, and knowledge of how a material will perform and feel like under the machine. All participants made decisions to add sub-task variations according to fabric type and personal preference to arrive at the same result. The following accounts not only demonstrate the decision-making happening prior to and during the task, but crucially also highlight how often they remark on possibly using unique approaches to them, showing a personal and subjective ability to handle the fabric, hinting at tacit or experiential skill.

During the experiments for this trial, a critical element that helped evidence skill in multiple ways lay in the restriction that no pressing iron could be used during task completion, forcing the participants to find a workaround to achieve a high standard despite not being able to iron halfway through the task. For example, fabricating the placket without pressing the fabric during construction led them to make use of different tools and techniques. In sub-task variation 4.3.2, participant 2 remarks on her decision to use pins to construct the placket: "[…] folding all the layers together and pinning... I use pins, although some seamstresses don't use pins. I use pins because it's keeping the fabric in place, especially if we're not using the iron." Based on the restriction, she decides to rely more heavily on pins than seamstresses might normally do. The pressing restriction also led them to different approaches to folding the placket edges. Participant 3 started off with an unpressed placket and later creased the edges by manipulating the fabric with her fingernails; participant 2 pre-pressed the edges of the inner and outer plackets prior to sewing activity and later used her fingernails to reinforce the crease; and participant 1 also decided to mix approaches and opted for a combination of adding stay-stitches to the placket edges to ease the fabric and folding with their fingernail. Participant 1 explains her choice as follows: "This is my stay-stitch [...], not everyone is doing this. When I do a stay-stitch, I don't need ironing on cotton fabric, [it] is very easy. I use my hands to fold [the] material and this stay-stitch helped me because it's keeping [a] straight line, that's why I'm not use[ing] the iron for this, depends on the fabric." This description of sub-task variation 4.1.1 gives accounts of multiple cognitive decision activities. Firstly, she decides to add a stay-stitch as a visual and physical tool to know exactly where to fold, while adding that this technique is something possibly unique to her. Secondly, she decides to do this as an *easy* alternative to ironing the fabric, knowing that it will yield quality results. And thirdly, she decides on this technique based on the cotton material and adds that using it depends on the fabric. All three approaches lead to the same result, but the order of subtasks added and extra tools used to complete the placket slightly changed depending on the techniques chosen.

In sub-task variation 2.3.1, participant 1 reinforces the placement line with additional stitches to make sure that the stitch line is not going to open accidentally when the seam allowance is cut into. She gives an oral account that introduces this action as something specific to her: "And I just do back stitches as well on every corner. It's not [that] everyone does this, but because we need to snip the middle..." The high level of skill and knowledge demonstrated, and decisions made prior to and during task performance were heavily influenced by previous experiences in handling similar fabric. The material properties therefore inform how the machinist deploys their tactile knowledge (Tallis, 2003). The speed and available functions of the sewing machine played an equally crucial role in discerning how to handle the material to maximum effect. In addition to the aforementioned factors, personal preferences play a huge role in choosing particular techniques. A combination of those factors then informs the distinctive approach chosen by each sewing machinist. This provided them with assurance to realise the task accurately and to a high standard.

In reviewing the data despite the experiential scope of this research, the research team clearly evidenced the existence of experiential skill or tacit skill in sewing machinists. What's more, it was able to capture and formalise the techniques to inform tooling design more broadly. The findings suggest that the manifold handling skills present in this small sample imply that for a tool to be useful, it needs to be flexible in its use and capable of accommodating many practitioners approaches to material handling and tool use. Mono-solutions and monotonous use are to be avoided when designing for flexibility and evolving skills in a human-centred Industry 5.0 context.

2.2.4.2 Low Accuracy versus High Qualitative Data

None of the five data sets captured the pupil movement of the participants with full accuracy. Rather, deviations were suspected as soon as the data was coded. Using the Method Stacking approach to data capture meant that the data was not discounted but used constructively by utilising the limitations of the physiological measures to guide the discussions during the reflective interviews. Participants noticed at various points that the focal tracker in the videos was off-target and not representative of the focus of their attention. Sewing tasks demand a high level of precision, focus, and accuracy, and deviations did not go unnoticed.

When constructing a placket, there are several critical steps demanding the sewing machinists' full attention. One such moment occurs when the placket is being cut open in a Y shape, with the very tips of the Y bordering directly on the box seam corners (sub-task 3.2). Participant 2 remarks: "On the screen, it looks like I was looking above the detail of the two corners, but I was actually really focusing my eyes on the two corners. Because with cutting like this, you really have to be precise; the cut needs to go straight to the stitch line. But you have to be careful that you do not cut over the thread. If I would cut over the thread that means I have to do it all over again, because once I started turning if there's no stitch line or if the thread is cut it just wouldn't work." Just a few moments later, around sub-task variation 4.3.2, she notices the discrepancy again: "Now, I'm again.... my eye, it looks like it's going above, but I was looking exactly where I was pinning, you want to be really precise when you're pinning this part. Because you have to cover the stitch line that you made the first time, the bottom and the top layer have to match." And another remark around 5.3.2 when she pins the pre-pressed placket to the stitch line, another crucial moment to ensure quality work: "I'm pinning in another way, so not from side to side, but more like along the stitch line. And just making sure that everything is folded neatly so there's no puckering and all the layers are hidden underneath the placket. I'm also using the pin to hide the little corner. It's just for the precision of it. Yeah, I think the line of focus is probably about a centimetre off or a couple of centimetres off. With details like this, you really have to focus, and especially with pinning, you look where you're pinning." In all of these instances, participant 2 stresses that the focal tracker did not capture her attention to detail.

She highlights this concisely by explaining the intentions behind her heightened focal attention to particular steps in the process. Adding to the complexity of processes when creating the placket, she describes visual and tactile inspection processes in great detail (*just making sure that everything is folded neatly so that there's no puckering*) and small moments of tool use to ensure every step is done with precision (*I'm also using the pin to hide the little corner. It's just for the precision of it*). Focus, precision, and constant visual inspection come up repeatedly in her assessment of the drivers of her own actions from moment to moment. At another crucial step, when a small triangle of the Y shape is secured into the body of the placket, participant 1 remarks: "I think it's [a] little bit drifted because [according to the focal tracker] I'm looking [at the] material and not on the metal. Yes, I'm looking at the needle! Maybe because I have two different glass[es] on. I don't know."

Seeing the tracker malfunctioning in the video recording led to vivid responses by the participants, who were keen to rectify that what was shown on-screen was indeed inaccurate and not demonstrative of their skilful handling and attention to detail. This is significant as it demonstrates that physiological data capture might return data that is only seemingly accurate for people unfamiliar with the precision needed to execute the task. This demonstrated the positive effect of Method Stacking to overall study design. The research team assumes that if the focal point had been persistently accurate, participants would not have given such detailed accounts of their actions, motivations, and knowledge on how to handle the task. By choosing not to discount the data sets by means of their apparent shortcomings, the research team was able to co-construct knowledge with the participants and challenge their own confirmation bias towards a method's assumed accuracy. Method Stacking is therefore a useful research approach helping transdisciplinary teams to remain open and responsive to data interpretation and validation. By gradually stacking evidence, the team was able to collect rich feedback and insights.

2.2.5 Limitations

2.2.5.1 Eye-tracking and Prescription Glasses

Physiological data capture can be intrusive and create discomfort in some people. Most research in eye-tracking focuses on human–computer interactions, predominantly the capture of screen-based activity. Studies that situate eye-tracking in real-life scenarios are still sparse but have gained ground in recent years (Howell et al., 2023; Berni et al., 2022). Depending on the eye-tracking model, devices now come with specifically designed attachable prescription lenses that do not interfere with the focal tracker. In lieu of attachable prescription lenses, participants might struggle to execute tasks to the best of their abilities while wearing the eye-tracking devices only. All participants of this study relied on spectacles which challenged the SMI device considerably. Accurate and consistent capture of pupil movement was overall difficult to achieve, and it is presumed that such deviations arise from lens flares which are obstructing an unimpeded view of the pupil.

In that case, a balance must be found between ensuring as few anomalies as possible in the recording and maintaining the quality of the output. The eye-tracking performed overall more reliably when worn without prescription lenses. The two instances where correctives were removed – during participant 2's test and participant 3's second test – resulted in enhanced performance of the eye-tracking. This improvement was however hampered as it coincided with a reduction in stitch accuracy and general quality of the placket, as even minor deviations significantly impact sewing excellence. It is therefore advised for future tests to utilise eye-tracking devices that are purposefully manufactured to include options for prescription lenses. Lens flares and irregularities might still occur in data sets captured with inbuilt prescription options due to changing light situations and complexity of the physical task. Method Stacking has proven to be a useful approach to utilise the potential limitations constructively. It is therefore advised to be carried out even if the eye-tracking accuracy is improved upon.

2.2.5.2 Triangulation with Other Physiological Measures

It is further suggested to triangulate the eye-tracking with brainwave monitors/EEGs for any future research in this area. Participant feedback revealed that a sense of accomplishment is one of the main indicators for a successful project, activating a personal reward system crucial for creative learning. This activation cannot be evidenced through eye-tracking however. Related to the study of cognitive reward is the nascent field of Design Neurocognition that investigates design cognition and creativity during design activities (Balters et al., 2023; Gero & Milovanovic, 2020;

Vieira et al., 2019), mainly using functional magnetic resonance imaging (fMRI), functional near-infrared spectroscopy (fNIRS), and electroencephalograms (EEGs).

For instance, investigations employing fMRI to examine creative problem-solving activities unveiled the essential involvement of the brain's reward network during moments of insight or eureka/AHA! Experiences (Tik et al., 2018). The experience of reward upon task completion contributes to reinforced learning, thereby facilitating memory consolidation. Complementary studies within this domain utilising electroencephalograms (EEGs) (Benjaboonyazit, 2016; Sandkühler & Bhattacharya, 2008) propose that successful problem-solving and the occurrence of AHA! Moments are contingent on participants overcoming their Psychological Inertia (Altshuller et al., 1998), mental impasse, or functional fixedness. Significantly, research indicates that inertia tends to be more pronounced with increasing expertise and skill. This presents intriguing avenues for further exploration into meaningful and rewarding work, as well as its implications for HMI and human–robot interactions (HRI).

2.3 CONCLUSION

This study has delivered perspectives on ways to test and ultimately increase collaboration capability between Human Factors and Fashion Practice Research to innovate within emerging manufacturing processes amidst a growing skills shortage in the UK. It has evidenced and presented the expert skills possessed by sewing machinists, highlighting the manifold experiential skills developed by users of sewing machines. Understanding the personal handling styles of sewing machinists in more detail enabled the researchers to draw directional conclusions that will help inform future studies in support of the design of new tooling for fashion garment manufacturing.

Acknowledging increasingly integrated manufacturing business models, this study has delivered the beginning of a collaborative protocol for the study and design of new tooling with multiple designer/users and utilities in mind. Expert practitioners' sensitivity to material, dexterous handling, and creative problem-solving can be leveraged in the design process to enable the continuous emergence of skill through new cobotic tooling. As such, new cobotic sewing systems need to be understood as a site for multi-technique and multi-skill development.

To date, robotics arms have yet to find adequate use cases in smaller business operations in the fashion sector. Similar to other large industries, robotic pick-and-place operations are by far the most deployed within fashion manufacturing, particularly along large offshore assembly lines. Given the general absence of such large-scale fashion manufacturing in the UK, it is no surprise that the sector is predominantly comprised of SME and micro businesses, which often encounter challenges in automating parts of their production.

Collaborative robotic processes promise to assist SMEs in gaining more creative autonomy for on-demand and co-located manufacturing processes (Postlethwaite et al., 2022). This suggests that developing the systems needed to modernise the UK's fashion manufacturing sector and populating it with smart, agile, and collaborative robotic systems can ease pressure on recruitment and facilitate more dynamic and creatively rewarding work during production processes. There is huge potential for future studies in this area by developing systems that capture, apply, and develop transferable skills. This will support a rejuvenation of the job market creating opportunities that can inspire and entice a young workforce to enter into what can be a dynamic field.

It is important, however, to find a reasonable balance. A high level of skill was found to lead to excessive reliance on familiar techniques and processes in the creative field, referred to as design fixation (Crilly, 2018), which can often hinder creative practice. New research in the manufacturing sector looking at Operator 4.0 models reports a similar impasse called Manufacturing Fixation in Design (MFD) (Fillingim and Feldhausen, 2023; Bracken Brennan et al., 2021). It is also known that risk taking, which is an elemental factor in creating, can be hampered by rigid frameworks and excessive definition (Petreca, 2017). Aligning the development of new tooling to rewarding activity that is neither too restricted nor creating excessive cognitive load is therefore a critical next step in design research with Human Factors to address how designing new cobotic tools can help facilitate and mediate creative process.

ACKNOWLEDGEMENT

The authors would like to acknowledge Dr. Sarah Fletcher and Dr. Iveta Eimontaite at Cranfield University for their support in the continuing development of this transdisciplinary methodological approach. This research was made possible through funding from ASPECT (A SHAPE Platform for Entrepreneurship, Commercialisation and Transformation).

REFERENCES

Ahmad, A., Javed, I., Abrar, U., Ahmad, A., Jaffri, N. R., Hussain, A. (2021). Investigation of ergonomic working conditions of sewing and cutting machine operators of clothing industry. *Industria Textila*, 72(3), 309–314. doi:10.35530/IT.072.03.1723.

Altshuller, G., Zlotin, B., Zusman, A., Philatov, V. (1999). ARIZ. In: *Tools of Classical TRIZ*. Michigan: Ideation International Inc., Chapter 2, pp. 20–68.

Atkinson, D. S. (2022). *Tailoring digital touch: An ethnography of designers' touch practices during garment prototyping and the potential for their digitisation.* Doctoral thesis. UCL (University College London). Available at: https://discovery.ucl.ac.uk/id/eprint/10153438/ (Accessed 10 June 2024)

Balters, S., Weinstein, T., Mayseless, N., Auernhammer, J., Hawthorne, G., Steinert, M., Meinel, C., Leifer, L. J., Reiss, A. L. (2023). Design science and neuroscience: A systematic review of the emergent field of design neurocognition. *Design Studies*, 84, 101148. doi:10.1016/J.DESTUD.2022.101148.

Benjaboonyazit, T. (2016). Triz based insight problem solving and brainwave analysis using EEG during aha! moment. *Proceedings of the MATRIZ TRIZfest 2016 International Conference*, 2016, Beijing, China. Knoxville, TN: The International TRIZ Association – MATRIZ.

Berni, A., Altavilla, S., Ruiz-Pastor, L., Nezzi, C., Borgianni, Y. (2022). Human behaviour and design creativity. An eye-tracking study to identify the most observed features in a physical prototype of a tiny house. *International Design Conference - Design 2022,* Dubrovnik, Croatia. doi:10.1017/pds.2022.86.

Blader, S., Barck-Hoist, V., Danielsson, S., Ferhm, E., Kalpamaa, M., Leijon, M., Lindh, M., Markhede, G. (1991). Neck and shoulder complaints ergonomics among sewing-machine operators. *Applied Ergonomics*, 22(4), 251–257.

Bracken Brennan, J., Miney, W. B., Simpson, T. W., Jablokow, K. W. (2021). Manufacturing fixation in design: Exploring the effects of manufacturing assumptions on design ideas. *International Design Engineering. Technical Conferences and Computers and Information in Engineering Conference, Virtual /Online Conference,* 2021. American Society of Mechanical Engineers. doi:10.1115/DETC2021-70361.

Brohi, S., Khokhar, R., Marriam, P., Rathor, A., Memon, A. R. (2022). Prevalence of symptoms of work-related musculoskeletal disorders and their associated factors: A cross-sectional survey of sewing machine operators in Sindh, Pakistan. *Work*, 73(2), 675–685. doi:10.3233/WOR-210620.

Chan, J., Janowitz, I., Lashuay, N., Stern, A., Fong, K., Harrison, R. (2002). Preventing musculoskeletal disorders in garment workers: Preliminary results regarding ergonomics risk factors and proposed interventions among sewing machine operators in the San Francisco Bay Area. *Applied Occupational and Environmental Hygiene*, 17(4), 247–253. doi:10.1080/10473220252826547.

Crilly, N. (2018). 'Fixation' and 'the pivot': Balancing persistence with flexibility in design and entrepreneurship. *International Journal of Design Creativity and Innovation*, 6(1–2), 52–65.

Das, S., Natarajan, S. (2022). A study on musculoskeletal disorders in garment industry. *Tekstilna Industrija*, 70(2), 61–66. doi:10.5937/tekstind2202061d.

Delleman, N. J., Dul, J. (1989). Ergonomic guidelines for adjustment and redesign of sewing machine workplaces. In: Haslegrave, C. M., Wilson, J. R., Corlett, E. N., Manenica, I. (eds.), *Work Design in Practice*. London: Taylor & Francis, pp. 155–160.

Design Council. (2018). *Designing a Future Economy - Developing Skills for Productivity and Innovation*. Edited and produced by the Design Council. Online Publication. Available at: designcouncil.org.uk/fileadmin/uploads/dc/Documents/Designing_a_future_economy18.pdf.

Fillingim, K. B., Feldhausen, T. (2023). Operator 4.0 for hybrid manufacturing. *Proceedings of the Design Society*, 3, pp. 2835–2844. Published online by Cambridge University Press. doi:10.1017/pds.2023.284.

Frayling, C. (1993). *Research in Art and Design*. London: Royal College of Art Research Papers.

Gaver, W. (2012). What should we expect from research through design? *Proceedings of the 2012 ACM Annual Conference on Human Factors in Computing Systems - CHI'12*. New York: Association for Computing Machinery, pp. 937–946. doi:10.1145/2207676.2208538.

Gero, J. S., Milovanovic, J. (2020). A framework for studying design thinking through measuring designers' minds, bodies and brains. *Design Science*, 6, e19. doi:10.1017/dsj.2020.15.

Gunn, W., Donovan, J. (2016). Design anthropology: An introduction. *Design and Anthropology*, pp. 1–16. doi:10.4324/9781315576572-5.

Harris, J., Begum, L., Vecchi, A. (2021). *Business of Fashion, Textiles & Technology: Mapping the UK Fashion, Textiles and Technology Ecosystem,*. London: University of the Arts London.

Hooper, L., Lack, M., Carney, L., Postlethwaite, S. (2022). *UK Textiles Manufacturing: Opportunities and Challenges for the UK and Midlands*. Industry Report. London: Loughborough University.

Howell, B., Jackson, A. R., Edwards, A. M., Kilbourn-Barber, K., Bliss, K., Morgan, A. P. (2023). Assessing Eye Gaze Patterns between Intermediate and Advanced Design Sketchers'. In *Proceedings of the International Conference on Engineering Design (ICED23)*, Bordeaux, France, 24–28 July 2023, pp. 24–28. doi:10.1017/pds.2023.66.

Johnson, T. L., Fletcher, S. R., Baker, W., Charles, R. L. (2019). How and why we need to capture tacit knowledge in manufacturing: Case studies of visual inspection. *Applied Ergonomics*, 74, 1–9. doi:10.1016/j.apergo.2018.07.016.

Koskinen, I., Zimmerman, J., Binder, T., Redstrom, J., Wensveen, S. (2011). *Design Research Through Practice: From the Lab, Field, and Showroom*. San Francisco, CA: Elsevier Science & Technology.

Krogh, P. G., Koskinen, I. (2022). How constructive design researchers drift: Four epistemologies. *Design Issues*, 38(2), 33–46.

Küçük, M., Birol, Ş. (2021). Development of competency analysis system for sewing operators in clothing industry. *Proceedings of the 4th International Conference on Modern Research in Engineering, Technology and Science*, Amsterdam, Netherlands.

Li, G., Haslegrave, C. M., Corlett, E. N. (1995). Factors affecting posture for machine sewing tasks. And the need for changes in sewing machine design. *Applied Ergonomics*, 26(1), 35–46. doi:10.1016/0003-6870(94)00005-j.

Petreca, B. B. (2017). Giving body to digital fashion tools. In: Broadhurst, S., Price, S. (eds.), *Digital Bodies: Creativity and Technology in the Arts and Humanities*. London: Palgrave Macmillan, pp. 191–204.

Polanyi, M. (1958). *Personal Knowledge: Towards a Post-Critical Philosophy*. Chicago, IL: University of Chicago Press.

Postlethwaite, S., Thiel, K., Atkinson, D. (2022). *Reshoring UK Garment Manufacturing with Automation. Recommendations for Government*. Research Report. KTN Made Smarter. Online Publication. Available at: https://iuk.ktn-uk.org/wp-content/uploads/2022/03/Reshoring-UK-Garment-Manufacturing-with-Automation-Thought-Leadership-Paper-final-2.pdf.

Ray, T. (2009). Rethinking Polanyi's concept of tacit knowledge: From personal knowing to imagined institutions. *Minerva*, 47(1), 75–92. doi:10.1007/s11024-009-9119-1.

Sandkühler, S., Bhattacharya, J. (2008). Deconstructing insight: EEG correlates of insightful problem solving. *PLoS One*, 3(1). doi.org/10.1371/journal.pone.0001459.

Sarder, B., Imrhan, S., Mandahawi, N. (2007). Ergonomic workplace evaluation of an Asian garment factory. *Journal of Human Ergology*, 35(1–2), 45–51.

Smith, E. A. (2001). The role of tacit and explicit knowledge in the workplace. *Journal of Knowledge Management*, 5(4), 311–321.

Smith, K. (2012). Sensing design and workmanship: The haptic skills of shoppers in eighteenth-century London. *Journal of Design History*, 25(1), 1–10.

Tallis, R. (2003). *The Hand: A Philosophical Inquiry into Human Being*. Edinburgh, Scotland: Edinburgh University.

The Alliance Project Team. (2015). *Repatriation of UK Textiles Manufacture*. Report. The Alliance Project. Online Publication. Available at https://ukft.s3.eu-west-1.amazonaws.com/wp-content/uploads/2018/05/13115441/Repatriation-of-UK-textile-manufacture-The-Alliance-Project-Report.pdf.

The Environmental Audit Committee. (2019). *Fixing Fashion: Clothing Consumption and Sustainability*. Sixteenth Report of Session 2017–19. London: House of Commons.

Thiel, K., Eimontaite, I. (2023). Method stacking. In: Rodekirchen, M., Pottinger, L., Briggs, A., Barron, A., Eseonu, T., Hall, S., Browne, A. L. (eds.), *Methods for Change Volume 2: Impactful Social Science Methodologies for 21st Century Problems*. Manchester: Aspect and the University of Manchester. https://e-space.mmu.ac.uk/634562/1/Method-Stacking-Report.pdf.

Thiel, K., Postlethwaite, S. (2023). Human-centric research of skills and decision-making capacity in fashion garment manufacturing to support robotic design tool development. In: Karwowski, W., Trzcielinski, S. (eds.), *Human Aspects of Advanced Manufacturing, AHFE (2023) International Conference, AHFE Open Access*, vol. 80. Online publication. AHFE International, Honolulu, HI. doi:10.54941/ahfe1003505.

Tik, M., Sladky, R., Luft, C. D. B., Willinger, D., Hoffmann, A., Banissy, M. J., Bhattacharya, J., Windischberger, C. (2018). Ultra-high-field fMRI insights on insight: Neural correlates of the aha!-moment. *Hum Brain Mapping,* 39(8), 3241–3252. E-Pub.

Future Employment Projections Sewing Machinist UK - Union Learn. (2023). https://www.unionlearn.org.uk/careers/sewing-machinists (Accessed 24 January 2023).

Vaughan, L. (2019). Designer/practitioner/researcher. In: Vaughan, L. (ed.), *Practice-Based Design Research.* London: Bloomsbury Visual Arts, pp. 9–17.

Vear, C. (Ed.). (2021). *The Routledge International Handbook of Practice-Based Research*, 1st ed. London: Routledge.

Vieira, S., Gero, J. S., Delmoral, J., Gattol, V., Fernandes, C., Parente, M., Fernandes, A. (2019). Understanding the design neurocognition of mechanical engineers when designing and problem-solving. *International Design Engineering Technical Conferences and Computers and Information in Engineering Conference*, California, USA.

West Yorkshire Combined Authority. (2021). *Let's Talk Real Skills. Engineering and Advanced Manufacturing Skills Plans.* Report Commissioned by Calderdale College. Online Publication.

3 Cultural Obstacles to Servitization in Japanese Manufacturing Industries

Keiko Toya

3.1 INTRODUCTION

One of the barriers to Servitization in the Japanese manufacturing sectors is said to be customers' unwillingness to pay-for-services. According to the results of a qualitative survey of Japanese manufacturing executives, unlike contract societies such as those in Western Europe, Japanese customers do not pay for services willingly (Toya et al., 2016). They point out that this is a major obstacle for Japanese manufacturers seeking to transform their products into services. Other studies have recognized that paid services are also an obstacle for Western manufacturers (Gebauer et al., 2005; Bititci et al., 2006; Malleret, 2006; Mo, 2012; Barquet et al., 2013). However, there was no clear evidence to suggest that this is a Japan-specific phenomenon. On the other hand, this argument from the management level of the Japanese manufacturing industry is very strong. In this study, we will clarify this point from the perspective of differences in national cultural backgrounds and the influence of national culture on corporate culture. As will be discussed later, Japan is a high-context culture that endorses tacit understanding, and at the same time, it is an uncertainty-averse culture. This culture, which has taken root in general society, permeates not only BtoC (Business to Consumer) business but also BtoB (Business to Business) business, and it is believed to influence the attitude and behaviour of manufacturing executives and employees toward their customers. In dealing with this problem, Japanese companies can consider measures other than the straightforward method of persuading customers to pay their bills.

3.2 THEORETICAL BACKGROUND

3.2.1 National and Corporate Culture Comparative Perspective

Hofstede's six-dimensional culture scale (Hofstede, 1983; Money et al., 1998; Roth, 1995; Bergiel et al., 2012) was developed by H. Hofstede in the 1960s for organizational development at IBM. It was developed by H. Hofstede in the 1960s for IBM's organizational development and has since been extended, validated, adapted, etc. in several papers, including Hofstede's own.

According to a 1983 paper, a highly uncertainty-avoidance culture is one that is anxious about uncertain and unknown conditions, prefers to be conservative and

DOI: 10.1201/9781003505327-3

homogeneous with their surroundings, and is dependent on others in authority and organizations they belong to. A high level of uncertainty avoidance culture tends to rely on their set rules and existing ways of doing things. It is an attempt to control as much of the uncertainties as possible. Hofstede (1984) uses one of the items of the questionnaire to measure this tendency. “Company rules should not be broken – even when the employee thinks it is in the company’s best interests”.

In east Asian countries such as Japan, South Korea, and Taiwan, cultures have high uncertainty avoidance (relatively high in Asian countries except in Singapore and Hong Kong), whereas Scandinavian countries have low uncertainty avoidance, and North America is in the middle.

There are also many studies from the perspective of high-context and low-context cultures (Hall, 1976; Korac-Kakabadse et al., 2001; Nguyen et al., 2007; Usunier and Roulin, 2010). Hall (1976) divided national cultures into high- or low-context cultures in order to understand the differences in communication. According to Hall (1976) and Hall and Hall (1990), “a high-context communication or message is one in which most of the information is either in the physical context or internalized in the person, while very little is in the coded, explicit, or transmitted part of the message”. On the contrary, a low-context communication is direct, precise, dramatic, open, and based on true intentions (Gudykunst et al., 1988). In high-context cultures, communication is less direct and more inclusive, and people read the unspoken truth behind what is said. High context is common to the craftsmanship contained in the culture of manufacturing, which is learning by observing the physical movements, attitudes, and actions of one’s predecessors rather than writing them down. In low-context cultures, communication occurs through clear, direct expressions of intent, and people do not have to guess at the true meaning behind them. In some cases, the same thing may be said repeatedly for clarification.

The relationship between high uncertainty avoidance, which is considered a characteristic of Japanese culture, and high-context culture appears to be an axis of opposition. In reality, however, the two coexist. High context is consistent with what community members understand in terms of uncertainty avoidance because it is a common tacit understanding within the same community. For example, in the case of the Japanese manufacturing industry, a common understanding exists as a high context between manufacturing employees and their business partners to avoid uncertainty in the manufacturing process and maintain the quality of the finished product. Therefore, detailed manuals are created for the manufacturing process and workplace rules are established to avoid uncertainty. Therefore, both seemingly contradictory cultures coexist in the Japanese manufacturing industry and have become entrenched as the manufacturing culture described next.

3.2.2 Relationship between the Nature of Goods and Culture

3.2.2.1 Goods and Service

According to the S-D Logic, a singular form service means process of resource integration for value co-creation, whereas plural form services are resources to be integrated for value co-creation in this process (Vargo and Lusch, 2004a, b, 2008, 2012). Though Servitization is not adding services, it includes the plural form services as intangible goods. The traditional service research identifies service features such as heterogeneity, indivisibility, and vanishing nature. Services have uncertainty inherent in the nature due to their intangibility. Moreover, the labour of service employees is emotional labour; they manage emotions; in other words, "they sell their emotions in exchange for wages" (Hochschild, 1983). Human emotions are unstable and are subject to internal conditions, such as physical condition, and external conditions, such as the environment. Compared to natural resources, uncertainty is inevitable because human beings, a resource of unstable quality, are invested as resources in the value creation process of services. Japanese people generally dislike this uncertainty. Typical examples of quality control include QC circles and ISO international standards. Therefore, the corporate culture of manufacturing firms is more likely to avoid uncertainty than that of other industries. The Japanese manufacturing industry has had a successful experience of excelling in product quality by maintaining its historical craftsmanship of "Monozukuri" culture (Fujimoto, 1999) while promoting automation and mechanization to improve productivity.

In summary, the national culture of uncertainty avoidance and high context influences the corporate culture of Monozukuri culture, which in turn influences Servitization via management and employees' behaviour and attitudes (Figure 3.1).

3.2.3 Japan's Unique View of Service

In Japan, the term "service" is often thought to refer to the hospitality industry, i.e., restaurants and hotels, within the tertiary sector. Inui and Matsukasa (2015) argue that hospitality services in Asia, particularly in Japan, have a different meaning than in the West. In the West, quality hospitality implies direct benefits, where companies can raise their rates and earn more profit, and employees can improve their personal skills and earn tips. In Japan, on the other hand, the effects of quality service are indirect. It satisfies employees' pride and contributes to the company through future repeat usage and brand enhancement. The payback to the individual employee is appreciation from the customer and a sense of professional self-efficacy. As such,

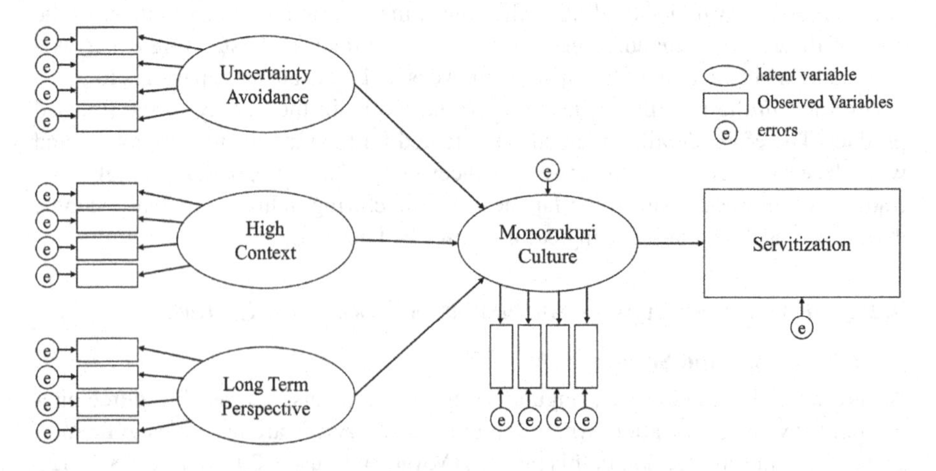

FIGURE 3.1 Conceptual model of corporate culture and Servitization.

many hospitality services have been provided free of charge in Japan as a sign of good faith (Hara et al., 2016; Goantara, 2019). Furthermore, some argue that charging a fee for something as expensive as hospitality diminishes its true value. This general view also affects employees of BtoB companies who receive services as individual consumers and applies to their work, whether consciously or unconsciously. For example, maintenance personnel at an industrial machinery manufacturer rush to the customer's site as soon as there is any problem with the machine. Even in the middle of the night, even on holidays, and even when repairs are not possible. Against this background, there is a certain rationale why Japanese manufacturers find it difficult to charge for services provided free of charge. In addition to being high context, the nuances of the term "service" are also considered to be an obstacle to charging for the service.

We have plotted the Japanese and Nordic cultures, goods and services, and manufacturing culture in the figure. As shown in Figure 3.2, Japanese culture has high uncertainty avoidance and high context, whereas Scandinavian culture has low uncertainty avoidance and context, and is located between Japanese and U.S. culture. On the other hand, services are low uncertainty avoidance because of the uncertainty inherent in their nature and high context in the sense that they are dependent on the customer's expectation level because it is difficult to evaluate quality based on external criteria. On the other hand, goods are high uncertainty avoidance and low context. Japanese manufacturing culture, while possessing the characteristics of goods, is a little closer to the high context side because of its philosophy of perfecting quality through craftsmanship, but tolerance for uncertainty is considered to be limited.

In the above, we discussed Japanese culture and its unique concept of service and Monozukuri culture in terms of its influence on the corporate culture of the manufacturing industry. In manufacturing Servitization, goods and services are integrated, and the goal is to co-create value with customers integrating their labour, knowledge, and skills. At this time, a manufacturing culture backed by high uncertainty avoidance and high context is considered a major obstacle to integration with unstable services. Adding to the complexity, in the case of BtoB, customers have similar

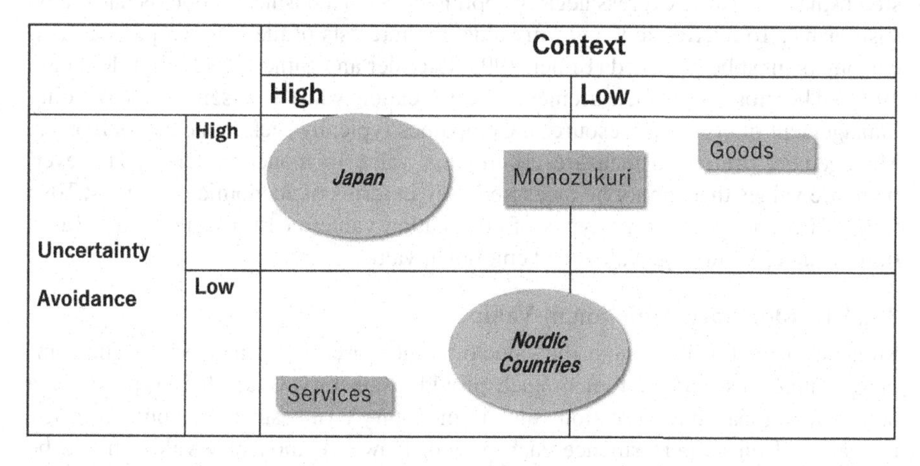

FIGURE 3.2 Uncertainty avoidance and high/low context related with Servitization.

corporate cultures, so Servitization will not be successful if only one side changes. When a manufacturing company engages in Servitization, which inherently involves uncertainty, the question is how to align with a high-context, high-uncertainty-avoidance culture with the agreement of all parties involved.

3.2.4 Three Dimension of the Co-creation Value

The concept of value co-creation, in which companies and stakeholders, including customers, provide resources and interactively create value, has been recognized in business administration since the 1980s (Toffler, 1980; Normann and Ramirez, 1993). Today, value co-creation is considered an important business process (Prahalad and Ramaswamy, 2004; Vargo and Lusch, 2004a, b, 2006, 2008). In the process of value co-creation (Vargo and Lusch, 2004a, b), resources are classified as either operant or operand. Operant resources are those that act upon other resources to create benefit, while operand resources are those on which an operation or act is performed to produce an effect. Value co-creation is an activity in which operand resources are transformed through an interactive process to create value through the mutual exchange of knowledge and technology between the company and its customers.

Measuring value co-creation is important in both academic research and business. In a matured economy, it is critical to redefine the way we think about value from the myopic relationship of buyer and seller to one based on a long-term relationship. Many previous researches indicate that business is transformed in this direction such as the long-term customer relationship management (e.g., Dholakia, 2001) and the royalty profit chain (Heskett et al., 1994). They argue that there is a virtuous cycle arising from loyalty among firms, employees, and customers in a long-term relationship. On the other hand, it is also true that the zero-sum game fighting over a certain share of value is existing among companies, employees, and customers. In a company that seeks to maximize the value of each exchange, someone has to lose out in order to compete with the customer. An unequal distribution of value leads the motivation to correct this inequality in the future. For example, customers may stop using the firm or express negative opinions. Such a business model is hard to be sustainable. To maximize long-term value, the interests of the firm, employees, and customers must be balanced (Bitner, 1993; Gremler and Bitner, 1992; Gremler et al., 1994). Therefore, new measurement of co-creation value is essential for ongoing management of corporate resources. Companies typically seek monetary outcomes. Management prefers to measure co-creation value in monetary terms. However, there are values that cannot be measured only in terms of economic outcomes; Toya (2014, 2016) proposes three types of co-creation value measures: monetary (functional) value, knowledge value, and emotional value.

3.2.4.1 Monetary (Functional) Value

Monetary (functional) value is the value of money or easily valued assets. The company, employees, and customers each provide monetary value. It is typical for a company to measure co-creation value in monetary terms, since the company itself is evaluated on its performance each period. However, monetary value should be measured over a longer period of time than the typical quarterly or semi-annual reporting cycle. Marketing goals to maximize market share are goals to maximize customer lifetime value (revenue earned from customers over their lifetime). Since monetary value is the result of the co-creation of value used over a lifetime, monetary value seems appropriate for measuring customer lifetime value. The customer lifetime value model is popular with firms that can track customer transactions on a customer-by-customer basis. However, the accuracy of predicting customer lifetime value is limited. Furthermore, by scoring customers in this way, firms overestimate the customers who are currently profitable and underestimate the customers who may be profitable in the future. Many researchers point out that monetary measures are insufficient and suggest the use of non-monetary measures such as ROQ in marketing (Rust et al., 1994), customer portfolio lifetime value in accounting (Johnson and Selnes, 2004), and balanced scorecards (Banker et al., 2000; Rust and Cooil, 1995; Rust et al., 1994, 2004, etc.).

3.2.4.2 Unmeasured Co-creation Value

Monetary value is an important part of co-creation value, but it does not represent all the value. We propose taking knowledge value and emotional value into consideration. Both values increase in the interaction between the provider and the receiver. Knowledge value is the value associated with the knowledge and skills cultivated among stakeholders. Emotional value is the value associated with the moods and perceptions of customers and employees. These non-monetary co-created values serve as resources in future co-creation, and part of them will ultimately create monetary value. The time period at which knowledge value and emotional value turn into monetary value varies and some part remains as it is.

The S-D logic emphasizes skill and knowledge, as well as the impact of that cognition on operand resources through knowledge and skills. In contrast, Normann (2001) explains that customer participation patterns can be categorized into three styles: physical, knowledge, and emotional. The physical participation is closely related to monetary value because it generates customers, or outside employees who co-produce the value for the companies. Knowledge participation is the provision and processing of information, skills, and advice, which is similar to knowledge value in our model. The emotional participation would become more important as the economic developments and maturities. Emotional participation is the provision of emotional energy. Technological developments have commoditized manufactured products, forcing them to compete emotional value based on the service instead of function. In this environment, emotional value becomes more important.

In the study of consumer behaviour, involvement means motivation and capability towards a goal. Involvement has both cognitive and affective aspects. Cognitive involvement is functional and practical, based on value creation through the use of a product or service. Emotional involvement mentally maintains and reinforces the ego.

3.2.4.3 Knowledge Value

Knowledge value is the accumulation of knowledge held by the co-creator that contributes to co-creation value. Peter and Olson (2002) identify two categories of knowledge: process knowledge and declared knowledge. Process knowledge relates to the service production process. Declared knowledge relates to events (episodic knowledge) and concepts (conceptual knowledge). Episodic knowledge is formed from experiences. Conceptual knowledge is more general and factual knowledge, e.g., "Brand A has feature B." In the context of consumer behaviour research, these types represent consumer knowledge. However, from the perspective of the service triangle framework, knowledge encompasses stakeholder knowledge about other stakeholders. As firms gain more knowledge about their customers' attributes, transactions, and other operating resources, they will be better able to design and deliver better co-production processes to their customers. Similarly, as firms gain knowledge about their employees, they will be able to offer more to co-creation processes and services, such as assigning appropriate tasks to employees or matching them with compatible customers.

3.2.4.4 Emotional Value

Emotions have positive and negative aspects (Watson and Tellegen, 1985; Bagozzi et al., 1999). Emotional value increases with stronger positive emotions and decreases with stronger negative emotions among participants in the service triangle. Sheth and Mittal (2004) identified eight types of emotions: fear, anger, joy, sadness, acceptance, disgust, expectation, and surprise. Because emotions in service strongly depend on the quality of service, surveys can be used to identify customer and employee emotions from these sets. In this study, emotion values are not assigned to firms, so our model includes emotion values for customer-to-firm, employee-to-firm, employee-to-customer, and customer-to-employee relationships. For the employee–customer relationship, we include short-term emotions such as excitement and joy. Long-term emotions such as trust are modelled in the customer-to-company and employee-to-company relationships.

3.3 QUALITATIVE SURVEY AND RESULT

All three companies have global operations, with Yamaha Motor having its origins in Japan and Nestlé Japan, a global company with its headquarters in Switzerland, having its origins in Japan. IKEA Japan, headquartered in Sweden, has a relatively short history in Japan, having arrived for the second time in 2002. The three companies listed here are all companies that have built business models that allow customers to enjoy some kind of co-creation value, whether financial or non-financial, when they are servitized.

Typically, when customers are unwilling to pay for a service, companies either absorb the cost through internal cost-cutting efforts and make the service free or pass it on to the product price (Kowalkowski and Ulaga, 2017). In addition, they may keep the product price to the distributor low, or instead of outsourcing part of the service process to the distributor, they may share the economic outcome of that part (e.g., sharing the costs incurred and sharing the revenue earned) with the distributor. This is the case of Yamaha Motor (Case 1).

In some cases, customers may be asked to perform part of the service process themselves in exchange for a lower product price. This is the case of IKEA (Case 2).

They may offer emotional value as a form of non-monetary co-creation value as a reward, recruit potential agents from among their customers, and outsource the work for free, with the role as a kind of privileged, branded position. The case of Nestle Japan (Case 3) is a case in point.

Before describing the individual cases, I will add an explanation of the above strategies. These strategies are not merely to maximize cost-effectiveness, but are also tactics aimed at encouraging customers to participate in the co-creation of value. When recruiting customers from the public to serve as sales agents, the compensation to customers is not always economic compensation; Toya (2016, 2020, 2022) states that the co-creation value of a service has three values: functional value, emotional value, and knowledge value; economic value is included in functional value and is a financial statement number. However, knowledge value and functional value, which are co-creation values that lead to the future, are not measured, although their existence is recognized, and they may mislead companies' sustainable strategies.

In the case of Yamaha Motor, the company shares the economic outcome, but in reality, the emotional value that the distributor marina derives from sharing and communicating the joy of marine leisure to its customers is significant; in the case of IKEA, customers gain the emotional value of self-efficacy, which is also true for Nestlé Japan. In the case of Nestlé Japan, the knowledge value that the company captures from customers through its ambassadors is important.

Case 1: Yamaha Motor Co., Ltd./Sea-Style

Company Name	Yamaha Motor Co., Ltd.
President	Yoshihiro Hidaka
Founded	July 1, 1955
Capital	86,100 million yen (as of June 30, 2023)
Sales	2,248.5 billion yen (Consolidated)
	517.0 billion yen (Marine Business)
Head Quarters	2500 Shingai, Iwata-shi, Shizuoka-ken, Japan
Employees (Consolidated)	52,554 (as of December 31, 2022)

Yamaha Motor Co., Ltd. runs a pleasure boat rental business in partnership with 140 marinas nationwide. After Japan's bubble economy burst in 1991, the pleasure boat ownership rate declined and the number of boat license holders (a license is required to drive a boat in Japan) aged, leaving the market stagnant Yamaha, the leading manufacturer, and its distributors, marinas in various regions, shared a sense of crisis that the market might disappear if nothing was done. Therefore, both parties worked together to create a new category of market, the rental boat market, for general customers to enjoy marine leisure activities. Until then, pleasure boats were considered to be only for the wealthy. With a subsidy from Yamaha, marinas buy pleasure boats for rental at almost half price and conduct the rental services; after three years of rental use, the boats can be sold as used boats. Marinas are like landowners who rent parking lots, and they only rent parking space to experienced boat owners, but the rental business requires a different kind of effort. Present a fishing or event program that might be of interest to consumers, and explain to members how to maintain their boats and describe nearby waters (e.g., shallow water areas to avoid and locations of fishing nets). Inexperienced drivers tend to have accidents when berthing their boats, so the company should pick them up and take over driving only for berthing. In addition, cleaning and fuelling the boats are other tasks for the marina. Even those who could not make a profit in the rental business due to poor sales activities can make a profit when they sell their used boats, and the business becomes profitable. Yamaha Motor will take on membership management, including the collection of membership fees, and will operate the reservation website. As of 2020, the number of members reached 20,000. This service revitalizes the pleasure boat market.

Yamaha Motor has redefined the roles of itself and its distributors and successfully converted them into services. The distributors' knowledge of local waters and customer needs are collected and shared among distributors nationwide. By creating and sharing economic value and knowledge value, the company is creating a new leisure category. Consumers are providing the resource of obtaining a boating license and doing the driving themselves, which in turn creates the emotional value of the enjoyment of marine leisure.

Case 2: IKEA

Company Name	Inter IKEA group
CEO	Jon Abrahamsson Ring
Founded	July 28, 1943
Group Equity	9,847 million EUR
Sales	44,600 million EUR
Head Quarters	Olof Palmestraat 1, 2616 LN Delft, Netherland (Inter IKEA Holding BV)
Employees (Consolidated)	231,000 (as of 2022)

Company Name	IKEA Japan K.K.
President	Petra Färe
Founded	July 8, 2002
Capital	8,605 million yen
Sales	95,418 million yen
Head Quarters	2-3-30, Hamacho, Funabashi-shi, Chiba-ken, Japan
Employees (Consolidated)	3,800 (as of August 31, 2021)

IKEA, founded in 1943 and originating from Sweden, has established a DIY process that allows customers to maintain and improve their own interiors and exteriors, and in some cases, build their own homes, unlike traditional furniture stores that sell assembled and finished products, based on the Scandinavian culture where DIY is popular. Since the customer's labour is invested not only in assembly but also in delivery, product prices are kept low. The customer identifies the materials needed for the work, brings them out of the stockroom, pays for them, brings them to his/her home, and does the work himself/herself. The product line ranges from simple furniture assembly to wall and flooring installation to larger remodelling projects. IKEA's huge retail stores have showrooms where customers can see how the work will be completed. IKEA once entered Japan in 1974 as a joint venture with Mitsui & Co. and Tokyu Department Stores, opening two stores, but withdrew in 1986. In a country like Japan, where the custom of DIY is not deeply rooted, the do-it-yourself style was not accepted. Additionally, the quality of IKEA furniture was not up to Japanese standards. Inconsistent sizes or insufficient numbers of screws in assembly furniture packages were a common occurrence, and consumers in Scandinavia were not particularly willing to make adjustments or procure them on their own. However, this caused dissatisfaction among Japanese consumers, who are accustomed to perfectly finished products. After addressing these issues, IKEA's 2006 re-entry into the Japanese market was a success, partly due to the DIY boom and the desire for fixed-price products after the burst of the bubble economy. IKEA stores are huge, and customers enjoy spending time there, browsing through showrooms and eating. IKEA saves costs by using its customers' labour, but at the same time creates value in the form of enjoyment for the customers themselves.

Case 3: Nestlé Japan K.K. Nescafe Ambassador

Company Name	Nestlé S.A.
CEO	U. Mark Schneider
Founded	1866
Capital	275 million Swiss franc (as of December 31, 2022)
Sales	94.4 billion Swiss franc (as of December 31, 2022)
Head Quarters	Avenue Nestlé 55, CH-1800 Vevey, Switzerland
Employees (Consolidated)	275,000

Company Name	Nestlé Japan Ltd.,
President	Tatsuhiko Fukatani
Founded	April 1913
Capital	4,000 million yen
Sales	No Data
Head Quarters	7-1-15 Gokodori, Chuo-ku, Kobe-shi, Hyogo-ken, Japan
Employees (Consolidated)	2,400

Nestlé Japan has named its employees "Nescafe Ambassadors," inviting them to voluntarily serve and manage coffee in the office. They are provided with coffee makers free of charge and are asked to keep the installation area clean, replenish coffee beans regularly, and collect fees from employees. The company receives orders from ambassadors and sends them coffee-related products for a fee. Each workplace is free to decide its own fees and collection methods. The company receives orders from ambassadors and sends them coffee-related products for a fee, either on a monthly or three-monthly subscription basis, with a minimum order of 6 boxes of 12 capsules per set, and after three orders, the company is free to cancel the order. In addition to house blend coffee, customers can choose from 20 different types, including café latte and matcha latte. Traditionally, this market has been a B-to-B category, with the vendor providing the coffee and managing the installation location. Because outsiders had access to the office, they had to go through a complex contracting process that included security checks and was expensive at the same time. In addition, these services were often introduced by large companies as entertainment for their guests, but did not meet the needs of the office workers or small- and medium-sized companies that worked there. Nestlé Japan's strategy focuses on meeting the needs of many office workers, such as the desire to drink good coffee in the office at a reasonable price (about 20 yen per cup), to stimulate office communication through coffee, and to connect with others. Ambassadors receive no monetary reward, but are motivated by fulfilling their need for self-approval by being useful to their colleagues at work. 500,000 people have been appointed as ambassadors in Japan by 2020. This system not only contributes to sales of coffee-related products (coffee beans, cream, cups, muddlers, etc.), but also greatly aids in the collection of knowledge about customer needs gained through the ambassadors. Based on this knowledge, the company has developed health food and matcha-related products, and has conducted test marketing through its ambassadors, leading to the launch of these products. Nestlé Japan, like IKEA, has realized cost savings by treating its customers as service providers, but also as co-developers in the product development process. This can be called knowledge value creation. Customers, in this case ambassadors, gain emotional value from the opportunity to communicate with colleagues in the office and feel a sense of satisfaction and accomplishment from their contribution. Office workers simultaneously receive the economic value of low-cost, tasty coffee.

3.4 FINDINGS

One of the obstacles to servitization in the manufacturing sector has been considered to be the difficulty of converting from free to paid services in Japan. The study shows that the influence of national and national-culture-influenced corporate culture can be seen in the servitization of the manufacturing industry. Specifically, as services involve uncertainty, servitization is thought to be driven by the tolerance of that uncertainty. The study showed cases where this uncertainty is absorbed in the form of agents who intervene between the customer and the end customer. The case also showed that services are based on long-term relationships with customers in terms of the profit structure, and that it is important to consider the value of knowledge and emotions

TABLE 3.1
Summary of Three Cases

Company/Service	Context	UAI	Business	Target Customer	Billing Method	Co-creation Value (Benefit for Stakeholders)
Yamaha Motor Co. Ltd./Sea-Style	Low	High	Membership-based pleasure boat rental business at 140 marinas nationwide	• Marinas willing to take on new service business challenges to avoid tapering off their pleasure boat business • Customers who want to enjoy a different kind of marine leisure using pleasure boats • Customers who want to enjoy fishing from their own boats on trips around the country	Annual membership fee and rental fee from consumers	– Yamaha: Boat sales, annual rental membership fees, fixed margins from marinas, maintenance of pleasure boat market, contribution to the green economy including reduction of CO_2 emissions – Marinas: Boat purchase subsidies from Yamaha, rental fees, sales of used boats, contribution to the green economy through reduction of CO_2 emissions, etc. – Consumers: Contribution to the green economy through a new lifestyle of marine leisure using pleasure boats, availability of rental boats nationwide, reduction of CO_2 emissions, etc.
IKEA	Low	High	Selling furniture and interior components at low prices	Consumers who want to enjoy DIY	Extra service such as delivery or assembly, customers pay extra service fee	• Customers do DIY to create the ideal home they prefer • IKEA can offer components at low prices because customers carry and assemble components • Targeted customers who are willing to do DIY
Nestlé Japan Ltd./ Nescafé Ambassador	Low	High	Providing office coffee by *lending* coffee makers free of charge Users can enjoy low-cost, authentic coffee with their colleagues in the office under their own management	Employees who want to *contribute* to a place that stimulates communication in the office Employees who want to enjoy good coffee in the office at a low price Employees who want to enjoy a good cup of coffee without having to leave their work place	Coffee makers are lent free of charge; instant or capsules are purchased	• Nestlé Japan: Increased sales of office coffee products, reduced advertising and sales personnel cost • Customers: Self-efficacy by contribution to activation of office communication • Low-cost, authentic coffee in the office

that contribute to business continuity in the long term, rather than measuring business success only in terms of short-term economic value. These can be seen as examples of what might be called the Asian varieties of service capitalism (Jones and Ström, 2018), indicating that servitization in Japan may not work as it does in the West. At the same time, the findings also show how servitization can contribute to the transition to a green economy. A long-term perspective on service activities, such as sharing, circular economy, and more efficient use of services in industrial processes, is beneficial for both business development and the green economy (Ström, 2020) (Table 3.1).

3.5 CONCLUSION

In summary, the academic contribution of this study is that it has examined the ways in which value can be created in servitization in relation to the characteristics of services and culture, and it has obtained directions for forms of servitization other than billing. In practical terms, the study also suggested that manufacturers should consider different ways of valuing value rather than rushing into service billing, taking into account the strong uncertainty avoidance orientation of their own companies and their customers. This could include the long-term experience of Japanese companies in creating various service solutions, such as product-package combinations (Bramklev and Ström, 2011). It is important to understand the value that customers seek, combine this with a long-term relationship-building approach, create a mechanism to involve customers in the co-creation of value, and find a value-creation model that fits into this mechanism. Companies have been pursuing short-term sales and profits based on shareholder capitalism for the past few decades, but in Japan, which has a different cultural background than the West, the finding that it is also effective to seek the path of servitization, which aims to improve non-financial value through co-creation, is a significant contribution from both a practical and academic perspective. We believe that this is a significant contribution, both practically and academically.

ACKNOWLEDGEMENT

This work was supported by the Servitization Consortium (https://unit.aist.go.jp/harc/servitization-conso/conso_overview.html).

REFERENCES

Bagozzi, R. P., Gopinath, M., and Nyer, P. U. 1999. The role of emotions in marketing. *Journal of the Academy of Marketing Science*. 27(2): 184–206.

Banker, R., Potter, G., and Srinivasan, D. 2000. An empirical investigation of an incentive plan that includes nonfinancial performance measures. *The Accounting Review*. 75(1): 65–92.

Barquet, A. P. B., de Oliveira, M. G., Amigo, C. R., Cunha, V. P., and Rozenfeld, H. 2013. Employing the business model concept to support the adoption of product-service systems (PSS). *Industrial Marketing Management*. 42(5): 693–704.

Bergiel, E. B., Bergiel, B. J., and Upson, J. W. 2012. Revisiting Hofstede's dimensions: Examining the cultural convergence of the United States and Japan. *American Journal of Management*. 12(1): 69–79.

Bititci, U. S., Mendibil, K., Nudurupati, S., Garengo, P., and Turner, T. 2006. Dynamics of performance measurement and organisational culture. *International Journal of Operations and Production Management*. 26(12): 1325–1350.

Bitner, M. J. 1993. Managing the evidence of service. In: Scheuing, E. E., and Christopher, W. F. (eds.), *The Service Quality Handbook* (pp. 358–370). New York: Amacom.

Bramklev, C., and Ström, P. 2011. A conceptualization of the product/service interface: Case of the packaging industry in Japan. *Journal of Service Science Research*. 3: 21–48.

Dholakia, P. 2001. Customer relationship management: The three myths of financial services CRM. *Financial Services Marketing*. 3(2): 40–40.

Fujimoto, T. 1999. *The Evolution of a Manufacturing System at Toyota*. Oxford: Oxford University Press.

Gebauer, H., Fleisch, E., and Friedli, T. 2005. Overcoming the service paradox in manufacturing companies. *European Management Journal*. 23(1): 14–26.

Goantara, L. O. 2019. *Analysis on Japanese Hospitality Spirit of Omotenashi: Does It Work in Other Countries? A Case of Its Implementation in Indonesia*. Master's thesis, Lund: Lund University.

Gremler, D., and Jo Bitner, M. 1992. Classifying service encounter satisfaction across industries. *Marketing Theory and Applications*. 3: 111–118.

Gremler, D. D., Jo Bitner, M., and Evans, K. R. 1994. The internal service encounter. *International Journal of Service Industry Management*. 5(2): 34–56.

Gudykunst, W. B., Ting-Toomey, S., and Chua, E. 1988. *Culture and Interpersonal Communication*. Thousand Oaks, CA: Sage Publications, Inc.

Hall, E. T. 1976. *Beyond Culture*. Hoboken, NJ: Anchor.

Hall, E. T., and Hall, M. R. 1990. *Understanding Cultural Differences*. Hoboken, NJ: Intercultural Press.

Hara, Y., Maegawa, Y., and Yamauchi, Y. 2016. Sustainability and scalability in Japanese creative services. In: Kwan, S., Spohrer, J., and Sawatani, Y. (eds.), *Global Perspectives on Service Science: Japan. Service Science: Research and Innovations in the Service Economy* (pp. 159–172). New York: Springer.

Heskett, L., Jones, T. O., Loveman, G. W., et al. 1994. Putting the service-profit chain to work. *Harvard Business Review*. 72: 164–174.

Hochschild, A. 1983. Comment on Kemper's "Social constructionist and positivist approaches to the sociology of emotions". *American Journal of Sociology*. 89(2): 432–434.

Hofstede, G. 1983. The cultural relativity of organizational practices and theories. *Journal of International Business Studies*. 14: 75–89.

Hofstede, G. 1984. *Culture's Consequences: International Differences in Work-Related Values* (Vol. 5). Thousand Oaks, CA: Sage.

Inui, H., and Matsukasa, H. 2015. A study on the differences between the hospitality offered in Europe and the US with that of Asia. *Journal of Industry and Management of Industrial Management Institute*. 47: 1–13.

Johnson, M. D., and Selnes, F. 2004. Customer portfolio management: Toward a dynamic theory of exchange relationships. *Journal of Marketing*. 68(2): 1–17.

Jones, A., and Ström, P. 2018. Asian varieties of service capitalism? *Geoforum*. 90: 119–129.

Korac-Kakabadse, N., Kouzmin, A., Korac-Kakabadse, A., and Savery, L. 2001. Low- and high-context communication patterns: Towards mapping cross-cultural encounters. *Cross Cultural Management: An International Journal*. 8(2): 3–24.

Kowalkowski, C., and Ulaga, W. 2017. *Service Strategy in Action: A Practical Guide for Growing Your B2B Service and Solution Business*. Fontainebleau: Service Strategy Press.

Malleret, V. 2006. Value creation through service offers. *European Management Journal*. 24(1): 106–116.

Mo, J. 2012. Performance assessment of product service system from system architecture perspectives. *Advances in Decision Sciences*. 12: 1–19.

Money, R. B., Gilly, M. C., and Graham, J. L. 1998. Explorations of national culture and word-of-mouth referral behavior in the purchase of industrial services in the United States and Japan. *Journal of Marketing*. 62(4): 76–87.

Nguyen, A., Heeler, R. M., and Taran, Z. 2007. High-low context cultures and price-ending practices. *Journal of Product and Brand Management*. 16(3): 206–214.

Normann, R. 2001. *Reframing Business: When the Map Changes the Landscape*. Hoboken, NJ: John Wiley & Sons.

Normann, R., and Ramirez, R. 1993. Designing interactive strategy: From value chain to value constellation. *Harvard Business Review*. 71(7/8): 65–77.

Peter, P. J., and Olson, J. C. 2002. *Consumer Behavior and Marketing Strategy* (6th ed.). New York: McGraw Hill.

Prahalad, C. K., and Ramaswamy, V. 2004. Co-creation experiences: The next practice in value creation. *Journal of Interactive Marketing*. 18(3): 5–14.

Roth, M. S. 1995. The effects of culture and socioeconomics on the performance of global brand image strategies. *Journal of Marketing Research*. 32(2): 163–175.

Rust, R. T., and Cooil, B. 1994. Reliability measures for qualitative data: Theory and implications. *Journal of Marketing Research*. 31(1): 1–14.

Rust, R. T., Lemon, K. N., and Zeithaml, V. A. 2004. Return on marketing: Using customer equity to focus marketing strategy. *Journal of Marketing, 68*(1): 109–127.

Rust, R. T., Zahorik, A. J., and Keiningham, T. L. 1995. Return on quality (ROQ): Making service quality financially accountable. *Journal of Marketing*. 59(2): 58–70. https://doi.org/10.1177/002224299505900205.

Sheth, J. N., and Mittal, B. 2004. *Customer Behavior: A Managerial Perspective* (2nd ed.). Cincinnati, OH: South-Western College Publishing.

Ström, P. 2020. *The European Services Sector and the Green Transition*. Brusseles, Belgium: European Parliament, Directorate-General for Internal Policies.

Toffler, A. 1980. *The Third Wave*. New York: Morrow.

Toya, K. 2014. A study of structure of co-created value in service. *Japan Marketing Journal*. 131: 32–45. https://doi.org/10.7222/marketing.2014.003.

Toya, K. 2016. Measurement of the value co-creation -FKE value model. *Serviceology*. 3(2): 32–35. https://doi.org/10.24464/serviceology.3.2_32

Toya, K. 2020. Co-creation value indicators required in a service-oriented society. *MBS Review*. 16: 67–74.

Toya, K. 2022. Co-creation value indicators required in a service-oriented society. *International conference RESER2022,* Paris.

Toya, K., Watanabe, K., Tanno, S., and Mochimaru, M. 2016. Internal and external obstacles of servitization in Japanese major manufactures. *Spring Servitization Conference 2016*, Manchester.

Usunier, J.-C., and Roulin, N. 2010. The influence of high- and low-context communication styles on the design, content, and language of business-to-business web sites. *The Journal of Business Communication (1973)*. 47(2): 189–227.

Vargo, S. L., and Lusch, R. F. 2004a. Evolving to a new dominant logic for marketing. *Journal of Marketing*. 68(1): 1–17.

Vargo, S. L., and Lusch, R. F. 2004b. The four service marketing myths: Remnants of a goods-based manufacturing model. *Journal of Service Research*. 6(4): 324–335.

Vargo, S. L., and Lusch, R. F. 2006. Service-dominant logic: What it is, what it is not, what it might be. The service dominant logic of marketing: Dialog debate and directions. *Journal of the Academy of Marketing Science*. 6: 281–288.

Vargo, S. L., and Lusch, R. F. 2008. Why “service”? *Journal of the Academy of Marketing Science*. 36: 25–38.

Vargo, S. L., and Lusch, R. F. 2012. The nature and understanding of value: A service-dominant logic perspective. In: Vargo, S. L., and Lusch, R. F. (eds.), *Special Issue-toward a Better Understanding of the Role of Value in Markets and Marketing* (pp. 1–12). Bingley: Emerald Group Publishing Limited.

Watson, D., and Tellegen, A. 1985. Toward a consensual structure of mood. *Psychological Bulletin*. 98(2): 219–219.

4 Obsolescence Risk Management for Long-Life Defense Projects

Ceren Karagöz Katı and Esra Dinler

4.1 INTRODUCTION

The inability to get parts from their original suppliers due to the end of their product life cycles is known as obsolescence. Obsolescence issues arise when components are missing from systems that require maintenance. Defense projects usually have such an extensive, intricate, and expensive scope. Because of this project complexity, businesses require an organized approach to address obsolescence issues. Large-scale defense sector companies' spare parts are part of the product life cycle. One of the most important performance metrics for a project is the operational condition of the vehicles and equipment. Large-scale defense sector projects will always result in data being unavailable due to the complexity and diversity of the data that has to be examined. As stated in the signed contract, companies must supply customers and end users with defective or worn-out parts during the guarantee period. On the other hand, inaccurate demand forecasting could cause a significant inventory risk or unavailability after the warranty period, which could result in money losses. The cost of replacement parts makes up a significant portion of the life cycle costs of products. For example, the value of spare parts used annually by machinery that may have a 30-year lifespan is close to 2.5% of the machinery's original purchase price [1]. In fact, it's important to forecast and anticipate the parts that will fail and when. Furthermore, the contract conditions and the scope of the project may have an impact on the selection criteria. Establishing the selection criteria for spare components that must be retained as a backup after sales is essential to efficiently controlling obsolescence risks. Decision-Makers (DM) frequently take into account many factors, such as the lead time, cost, and failure rate of the item; additionally, they may take into account the required complex engineering skills in order for the part to be ready for use; and the necessity of an export license for parts imported from abroad. It becomes a multi-criteria decision problem since the evaluation of several factors is a part of the spare parts decision-making process. Many strategies have been put forth in the literature to help with spare parts management in determining the ideal stock level. Many researchers have had the opportunity to work in this field due to

DOI: 10.1201/9781003505327-4

the variety of characteristics a project or company exhibits. Operational research in spare parts management has been reviewed by Hu et al. [2]. Demand forecasting, supply chain optimization, and spare part classification are all covered in this article. Rojo et al. [3] evaluated the risk of components in a bill of materials for a product that might impair system maintenance. According to the study, DM should focus only on the most crucial aspects by evaluating crucial factors for every aspect of the risk assessment process and eliminating any remaining aspects from their list. The article emphasizes the importance of proactive management of obsolescence risk in product development and highlights the benefits of integrating obsolescence management into the overall product life cycle management strategy. This article highlighted that by adopting best practices for obsolescence risk assessment and management, companies could reduce costs, improve product quality and reliability, and increase their competitiveness in the marketplace. Braglia et al. [4] focus on implementing a multi-feature classification method for managing spare parts inventory. The complexity and efficiency of spare parts inventory management are the primary concerns addressed in the article. Researchers suggest using a multi-feature classification method as a practical approach to managing spare parts inventory. This method separates the parts into different groups, considering their qualities and requirements. The article provides a detailed description of the multi-feature classification method and proposes a model for its implementation. This model considers the various characteristics of spare parts in the inventory to facilitate the classification process. The ultimate goal is to gain more effective and optimized control over spare parts management, including stock levels, order lead times, and inventory costs. According to Auweraer et al. [5], data from the current system may affect the process of creating demand. Supçiller and Çapraz [6] used TOPSIS (Technique for Order Preference by Similarity to Ideal Solutions) and AHP (Analytical Hierarchy Process) Multi-Criteria Decision-Making techniques to develop a solution to the supplier selection problem with multiple criteria. Dhakar et al. [7] contend that high-rate spare part estimation is possible with planned and periodic maintenance; however, a small quantity of safety stock is required for unforeseen failures. Using ABC and optimization techniques, Kasap et al. [8] investigated the identification of important spare parts for machinery repair. By taking into account the importance of parts as determined by the ABC method, order frequency, and service level constraints, they enhanced the demand forecasting method. Sandborn [9] focuses on managing supply chain risks during product design, particularly regarding the rapid technological obsolescence of components. The article emphasizes the critical role that obsolescence risk management plays in designing long-lasting and high-value products, particularly in the defense industry. The article discusses various strategies for managing obsolescence risks during product design, including strategic product life cycle management and stock-level planning. "Lifetime Buy Optimization" is one of the strategies discussed in this article. Because Lifetime buy costs play an important role, especially in producing high-cost and long-lasting products. Employing the economic order quantity formula, Ghare [10] investigated the number of failures under constant demand over time.

The defense industry is crucial in safeguarding a nation's security and sovereignty. It involves the conception, production, and maintenance of intricate and technologically advanced systems. The smooth operation of these systems is contingent upon uninterrupted and effective spare parts management. Therefore, managing spare parts is critical to maintaining sustainable operational excellence within the defense industry. Not only does it ensure operational continuity, but it also has a direct impact on a country's defense capacity. Moreover, spare parts management is necessary to maintain, repair, and modernize military equipment. Therefore, an effective spare parts management strategy is integral to a nation's defense strategy.

One of the main challenges of spare part management in the defense industry is managing the obsolescence of parts. The life cycle of military equipment is significantly longer than that of commercial equipment, and spare parts often become obsolete before the equipment reaches the end of its life cycle. Another challenge is that the defense industry faces unique supply chain challenges affecting spare part management. For example, military equipment is often manufactured by a range of domestic and international suppliers, making it challenging to manage the supply chain and ensure the timely delivery of spare parts. Moreover, military equipment is often deployed in remote locations, making it difficult to access spare parts on time.

The company where this study is carried out is a leading defense industry company with many projects. With its broad range of products and services, the company offers innovative and advanced solutions to its clients. However, the company has been facing a significant challenge with its spare part management system, which is not being done systematically to include all of its projects. The company, without a systematic approach to spare parts management, has utilized "the minimum 10% spare part inventory strategy" in past projects to ensure appropriate levels of spare part stocks and to facilitate rapid response to customer demands. One scenario in which the company is experiencing issues with its spare part management system is the production of armored vehicles for a client's military. The company has several projects for this client, each with unique spare part requirements. Despite this, the company's spare part management system does not consider all these projects, leading to several challenges. The first issue that arises is the lack of coordination between different projects. For instance, a spare part required for one project may not be considered for another project, leading to duplication of efforts and wastage of resources. This increases the production cost and results in a loss of time as the team has to repeat the process. The second issue is related to the cost of spare parts. With multiple projects underway, the company has to maintain an extensive inventory of spare parts. However, the company often orders spare parts in excess, wasting resources due to the lack of a systematic spare part management system. Additionally, when the company does not have the required spare parts in its inventory, it has to order them from external suppliers, which increases the cost of production. Moreover, the company also faces obsolescence management problems due to its spare part management system's inefficiencies. Some spare parts may become obsolete over time, and the company may not be aware of this until it is too late. In such cases, the company has to spend additional resources to identify an alternative spare part, which delays the production process and increases the overall cost.

The purpose of this study is to lower the risk of a shortage by identifying the selection criteria for parts that must be retained as backup after sales for a medium-sized project of an armored vehicle manufacturer operating in Turkey. When deciding which spare parts to keep on hand, DM frequently takes into account factors like the part's lead time, cost, failure rate, need for an export license for parts imported from abroad, and need for complicated engineering knowledge before the part is operational. Spare-part decision-making is a multi-criteria decision problem because it involves evaluating several criteria. To solve this problem, the weights of the criteria are determined by the AHP method. Afterward, the importance coefficients of the parts are determined with TOPSIS and MOORA (Multi-Objective Optimization based on Ratio Analysis). These coefficients were used as parameters in the mathematical model. To manage component obsolescence risk in the after-sales stage of a company associated with the Turkish defense industry, a mathematical model has been constructed. Results like optimizing the process to determine the number of components needed to minimize the cost and maximize spare parts availability, as well as purchasing enough products to meet the requirements of the system during its predicted life cycle time, will be achieved through the use of the model suggested in the study. Manufacturers of armored vehicles in this industry can use the approach this study developed to lower the risk of unavailability and enhance their spare part decision-making processes.

4.2 OBSOLESCENCE RISK MANAGEMENT METHOD AND APPLICATION

To meet the needs for the estimated life cycle of the components, there must be an adequate supply of products in the system. Currently, the most significant components must be determined. When determining the essential components and the relative importance of these criteria, it is critical to take certain factors into account.

This study examines the management of obsolescence risk for components in a project at a defense industry company. The project consists of 5,678 components. Three criteria are used for assessment, and the importance coefficients for the components are established based on these criteria.

The weights of the criteria in the suggested method are determined using the AHP method, and the weights of the components are determined using the TOPSIS and MOORA methods by these criteria. The relative relevance levels of the criteria are determined using the AHP method. Then, based on the criteria, the component weights using the TOPSIS and MOORA methods are established to determine the presence of essential components. In this mathematical model, the objective function's parameters are derived from the components' importance coefficient. Mathematical model and solutions are obtained after establishing the importance coefficient of the components. Figure 4.1 provides a flow chart for the suggested method.

4.3 ASSESSMENT CRITERIA FOR COMPONENTS

The difficulty encountered by previous projects in the after-sales period has been examined, and as a result of this examination, which parts are needed more frequently and which parts have difficulty in procurement have been investigated. As a result, it has been observed that the three criteria below have significant importance in terms of spare parts management, and their importance coefficients are determined. In this study, three criteria were evaluated, and the explanations of these criteria are given below.

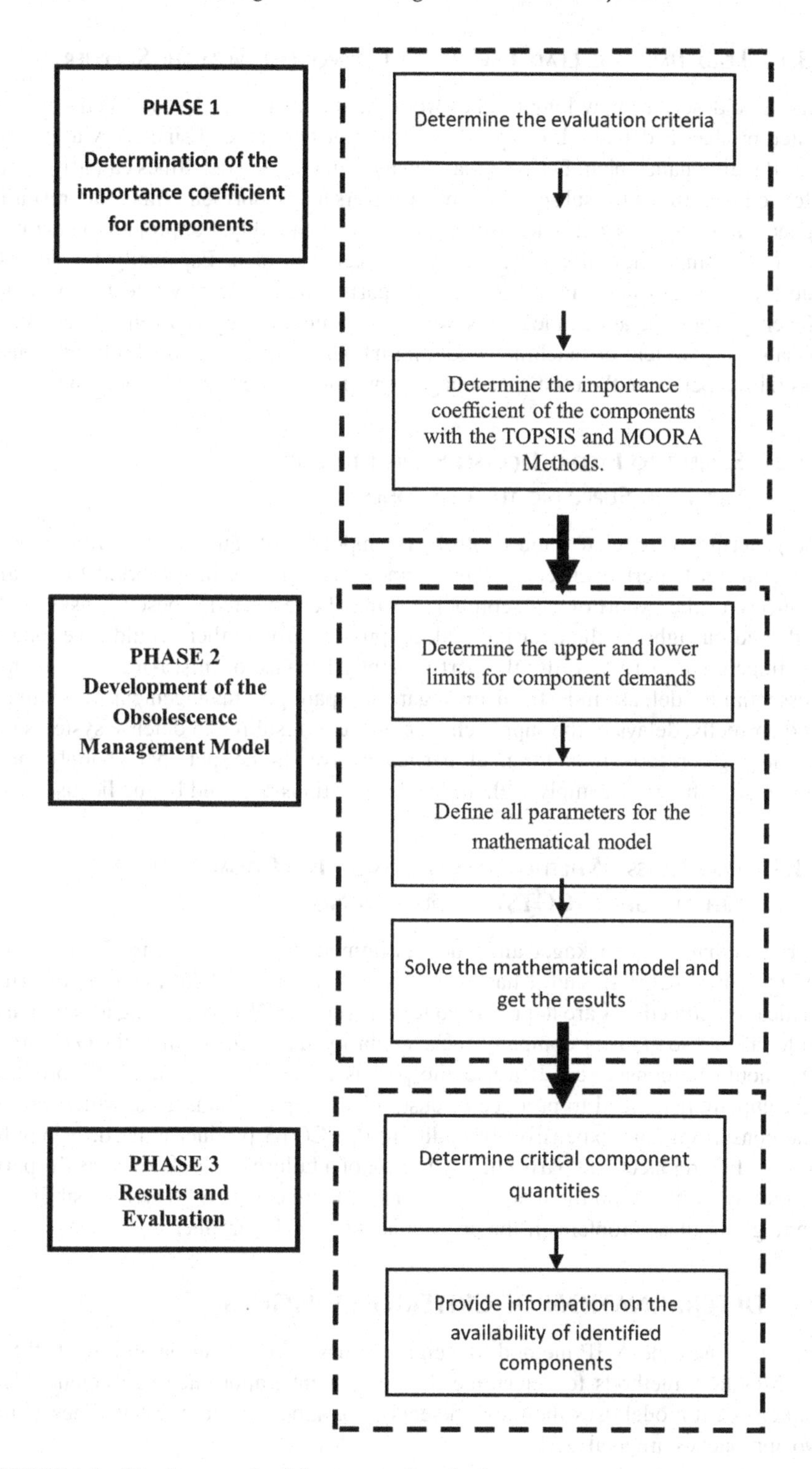

FIGURE 4.1 The framework of the proposed method.

4.3.1 Lead Time: The Lead Time of the Component from the Supplier

Lead time describes how long it takes from the moment an order is placed until the related product is delivered. For projects, lead times are crucial since they affect the overall maintenance plan. Delayed maintenance and missed deadlines can arise from extended lead times for sub-components. Conversely, a short lead time may result in higher inventory costs, extra inventory, and lower profitability. For projects, lead times are crucial since they affect the overall maintenance plan. Especially for defense industry firms, any delay in obtaining spare parts can significantly affect operational efficiency. These delays can lead to severe consequences, such as a vehicle becoming unusable or a system or machine becoming inoperable. The sustainability of a business thus depends on the efficient management of lead times for sub-components.

4.3.2 Subject to Export License: Subject to Export License in Supplying the Component

The government regulates and controls a component of international trade if it is subject to an export license. Certain components may not be exported to certain countries, or the export of such components may be restricted. These licenses could be denied outright, or the decision-making process around them could take longer. It is imperative that you order the part promptly because of this. An export license is essential for defense industry firms regarding spare part management. If not managed correctly, delays in the supply chain can occur, resulting in defense systems not functioning correctly. In addition, all parties involved in the spare parts supply chain must ensure that they comply with all legal regulations required by the license.

4.3.3 Part Class: Whether the Component Is a Commercial Off-the-Shelf (COTS) Product or Not

A product that is pre-packaged and ready for immediate sale is known as Commercial Off-The-Shelf (COTS), and it has a low order failure rate. Additional engineering verification procedures are not required for this product. When creating a spare parts list for a defense industry company, the availability and compatibility of COTS products should be considered. If a specific part is not available as a COTS product, the company may need to produce or customize the part themselves, which can be time-consuming and expensive. In addition, if a COTS product is used, it may be easier to find replacement parts quickly in case of a failure or breakdown, as the parts are readily available on the market. Conversely, there can be a greater possibility of running into these problems if the product is not a COTS product.

4.4 DETERMINATION OF CRITERION WEIGHTS

The study uses the AHP method to weight the assessment criteria and the TOPSIS and MOORA methods to determine the component importance coefficients. The mathematical model uses these coefficients as parameters, and the outcomes of the two approaches are analyzed.

Saaty introduced the multi-criteria decision-making method known as AHP in the 1970s [11]. The most crucial aspect of this method's application in the selection process is the evaluation of multiple qualitative and quantitative criteria. This approach is applied to several decision-making problems and has a broad range of uses. The first step is to identify the objective and the criteria that will relate to it. Pairwise comparison decision matrices are constructed to determine the relative importance of the criteria after they have been established. Pairwise comparison decision matrices are constructed to ascertain the relative importance of the criteria after they have been established. These matrices are generated by applying Saaty's nine-point importance scale [11]. By assessing the survey participants' or experts' opinions, this scale helps in determining the relative importance of different criteria.

The criteria weights in the AHP method were established by considering the opinions of three different DMs. Table 4.1 presents the decision-makers' judgments regarding the criteria.

The AHP method is used to determine the criteria weights, which are provided in Table 4.2, based on the evaluations listed in Table 4.1.

TABLE 4.1
The Judgments of the DM

	Export License	Lead Time	Class
	DM 1		
Export license	1	3	5
Lead time	1/3	1	7
Class	1/5	1/7	1
	DM 2		
Export license	1	3	4
Lead time	1/3	1	2
Class	1/4	1/2	1
	DM 3		
Export license	1	2	4
Lead time	1/2	1	1/3
Class	1/4	3	1

TABLE 4.2
The Criteria Weights

Criteria	Weights
Export license	0.619
Lead time	0.238
Class	0.143

4.5 DETERMINING THE IMPORTANCE COEFFICIENTS WITH THE TOPSIS METHOD

The TOPSIS (Technique for Order Preferences by Similarity to an Ideal Solution) method is a multi-criteria decision-making technique that ranks alternatives based on predetermined criteria. It was developed by Lai et al. [12]. Sorting the alternatives based on how near and far they are to the positive and negative ideals allows us to choose the best alternative.

The normalization of this matrix, which is essential for applying the TOPSIS approach, provides the relative importance of each criterion. After the decision matrix of alternatives is normalized and weighted using the relative weights of the AHP approach, both positive and negative ideal solutions are obtained and arranged in order. Next, the distance, both positive and negative, between each option and the ideal solution is calculated. The next step is to estimate the separation between each alternative and the ideal solution. Table 4.3 is produced following the classification of the alternatives. According to their large quantity, Table 4.3 contains a list of some of the components.

4.6 DETERMINING THE IMPORTANCE COEFFICIENTS WITH THE MOORA METHOD

In this study, the MOORA Significance Coefficient Approach is used to determine the importance of coefficients. The MOORA method was introduced to the literature in 2006 as a multi-criteria decision-making method by explaining its basic mathematical assumptions and creating and interpreting the results of its formulas in an application [13].

MOORA Importance Coefficient is an approach that facilitates the DM's evaluation and comparison between alternatives in a decision-making problem with the following aspects: It evaluates benefit- and cost-oriented criteria together, aspects of the relevant criteria can be easily determined, mathematical operations are quite easy to apply in the method, criterion importance coefficients are taken into account in the application of the method, and the situation of an alternative in cost–benefit analysis can be revealed concretely [13].

In this method, firstly the normalized decision matrix is obtained, and by multiplying each criterion in this matrix with the criterion weights which are determined by AHP, the weighted normalized decision matrix is obtained. Afterward, it is determined whether each of the criteria is benefit- or cost-oriented. Finally, in the weighted normalized decision matrix, the benefit–cost analysis value is calculated for all criteria by subtracting the sum of the values of the alternatives in the benefit-oriented criteria from the sum of the values in the cost-oriented criteria. These values obtained in this study are taken as the importance coefficient of the parts. Some of the obtained importance coefficients are presented in Table 4.4.

4.7 FORMULATION OF OBSOLESCENCE MANAGEMENT MODEL

This study proposes a mathematical model for managing obsolescence risk that, in the case that necessary components are unavailable, minimizes the overall risk. Purchasing components with high importance coefficients is prioritized in the model

TABLE 4.3
Component Data for TOPSIS Method

Sub Part Number	Export License (EL)	W_1*EL Norm	Lead Time (LT)	W_2*LT Norm	Class	W_3*Class Norm	S_i^+	S_i^-	Coefficient
804087	0	0.000	20	0.00066	0	0.00000	0.06678	0.000000	0.00000
808759	0	0.000	20	0.00066	1	0.00185	0.06671	0.00185	0.02695
805023-1	0	0.000	20	0.00066	1	0.00185	0.06671	0.00185	0.02695
800376	0	0.000	20	0.00066	1	0.00370	0.06668	0.00370	0.05250
803776	0	0.000	23	0.00076	0	0.00000	0.06676	0.00010	0.00149
802075	0	0.000	23	0.00076	1	0.00370	0.06666	0.00370	0.05254
113397-2	0	0.000	23	0.00076	1	0.00185	0.06669	0.00186	0.02691
803776	0	0.000	23	0.00076	0	0.00000	0.06676	0.00001	0.00149
803594	0	0.000	96	0.00318	0	0.00000	0.06632	0.00252	0.03658
805387	0	0.000	96	0.00318	1	0.00185	0.06624	0.00312	0.04502
804969	0	0.000	96	0.00318	1	0.00370	0.06622	0.00447	0.06326
805168	0	0.000	77	0.00255	0	0.00000	0.06643	0.00189	0.02764
801517	1	0.065	122	0.00404	1	0.00370	0.01017	0.06548	0.86556
805067	1	0.065	122	0.00404	1	0.00185	0.01034	0.06540	0.86352
812106	0	0.000	122	0.00404	1	0.00370	0.06608	0.00501	0.07044
807226	1	0.065	137	0.00454	0	0.00000	0.01036	0.06541	0.86331
801349	0	0.000	137	0.00454	1	0.00370	0.06600	0.00536	0.07505

TABLE 4.4
Component Data for MOORA Method

Sub Part Number	Export License (EL)	W_1*EL Norm	Lead Time (LT)	W_2*LT Norm	Class	W_3*Class Norm	Coefficient
804087	0	0.0000	20	0.000663	0	0.000000	0.000663
808759	0	0.0000	20	0.000663	1	0.001847	0.002510
805023-1	0	0.0000	20	0.000663	1	0.001847	0.002510
800376	0	0.0000	20	0.000663	1	0.003695	0.004358
803776	0	0.0000	23	0.000762	0	0.000000	0.000762
802075	0	0.0000	23	0.000762	1	0.003695	0.004457
113397-2	0	0.0000	23	0.000762	1	0.001847	0.002609
803776	0	0.0000	23	0.000762	0	0.000000	0.000762
803594	0	0.0000	96	0.003180	0	0.000000	0.003180
805387	0	0.0000	96	0.003180	1	0.001847	0.005027
804969	0	0.0000	96	0.003180	1	0.003695	0.006875
805168	0	0.0000	77	0.002551	0	0.000000	0.002551
801517	1	0.0653	122	0.004042	1	0.003695	0.073037
805067	1	0.0653	122	0.004042	1	0.001847	0.071189
812106	0	0.0000	122	0.004042	1	0.003695	0.007737
807226	1	0.0653	137	0.004539	0	0.000000	0.069839
801349	0	0.0000	137	0.004539	1	0.003695	0.008234

due to the determination of the objective function and constraints. The following are the sets, parameters, and decision variables:

4.7.1 Sets

I: the set of components, indexed by i

4.7.2 Parameters

c_i: the unit cost of component i
d_i: the amount determined to be available from the component i
B: the total budget
r_i: the importance coefficient of component i

4.7.3 Decision Variables

x_i: the quantity to be ordered for component i
u_i: the amount not available from component i
The obsolescence risk management model is given below:

$$\text{Minimize} \quad \sum_{i=1}^{I} r_i u_i \tag{4.1}$$

subject to

$$x_i + u_i = d_i, \forall i \tag{4.2}$$

$$\sum_{i=1}^{I} c_i x_i \leq B \tag{4.3}$$

$$x_i,\ u_i \geq 0 \text{ and integer}, \forall i \tag{4.4}$$

Equation (4.1) aims to reduce overall risk in the case that necessary components are unavailable. Constraints (4.2) determine the quantity of unavailable components or the amount of variation from the determined component amount. Constraint (4.3) guarantees that the overall budget is not exceeded. Constraints (4.4) are non-integrality constraint. By taking into account the amount of components determined, this model minimizes risk and establishes the appropriate amount of components to buy without going over the entire budget.

This study uses the fault records from the previous year to determine the amount that is available from the component (d_i). Ten percent of this sum is used to calculate the number of sales that fell within the project's parameters if there is no record of any component failed. The TOPSIS and MOORA methods yielded the values for the importance coefficients (r_i). The budget allotted for the current project is represented by the total budget parameter (B). Using the specified parameters and decision variables, a mathematical model for each of the project's 5,678 components is developed and the results are determined. The developed mathematical model was solved in CPLEX 22.1.1, and the results have been obtained by getting the algorithm in the background to find the objective function's optimal result.

4.8 RESULTS AND ANALYSIS

This study aims to optimize spare part risk management by examining spare part stocks in a large-scale defense industry company. The proposed method is aimed to find solutions to the following problems encountered in the company:

- Making enough product purchases to fulfill the needs for the anticipated lifetime of the system.
- Improving the procedure to calculate the quantity of parts required in order to reduce costs.
- Optimizing the supply of spare parts.

To achieve the target, the criteria for selecting spare parts by conducting a literature review and consulting with experts were determined. The AHP method is used to determine the relative importance of the criteria. By giving weights according to their importance in the decision-making process, it has been revealed that the "Export License" criterion has the highest weighted AHP score of 0.619. This highlights the vital role of export regulations in the defense industry. TOPSIS and MOORA methods are used to determine critical spare parts according to the determined criteria. Then, risk coefficients are assigned to each part according to the criteria, and comparisons are made according to these two methods.

In the developed mathematical model, since the objective function is formulated to minimize the total risk if the necessary components are not available, the formula will force to reduce the number of not available components since the amount not available from the component is a decision variable, and the parts with high risk are fixed values. In addition, while doing these, the budget constraint of the project is considered, and an optimization is made for both risky and costly parts.

Table 4.5 presents the components that were not available due to the TOPSIS approach mathematical model. The MOORA method's results are displayed in Table 4.6. Other components than those listed in these tables have had their needs satisfied. The needs of components with low significance coefficients and high unit prices are largely unmet, as can be seen from the results in Table 4.5. Table 4.6 displays comparable findings. Furthermore, Figure 4.2 presents the comparison of unit cost and importance coefficients of the components that are unavailable based on the TOPSIS technique, whereas Figure 4.3 presents the same comparison based on the MOORA method. It is obvious from both figures that the components with low significance coefficients exhibit density.

In this study, the importance coefficients are determined with two different multi-criteria decision-making methods, and mathematical model results are obtained with these two methods. According to the results (Tables 4.5 and 4.6), the number of components that could not be met was 76 in the MOORA method and 117 in the TOPSIS method. While it is observed that low-importance components are not met in both methods, it is seen that the number of unmet products is less in the MOORA method.

4.9 CONCLUSION

Spare parts management is a crucial issue in many industries and sectors. Spare part costs, especially in the defense industry, constitute a considerable part of the project costs. Appropriate spare parts management can significantly reduce operational costs, ensure production continuity, and increase customer satisfaction. In the defense sector, developing and maintaining products may include complex corporate processes and highly sophisticated technology. Managing and planning these systems can be difficult and complex, particularly in part supply or production situations.

In the defense industry, extensive business procedures are necessary for the development and maintenance of advanced equipment. The development and management of such systems are challenging and complex in scenarios such as production or component supply. There is a period called "end-of-life" that marks the conclusion of the product lifetime, which begins with product retirement and ends with the expiration of all service contracts. When a product reaches the end of its useful life and becomes obsolescent, remanufacturing it can be an alternative way to obtain spare parts. As a result, at each stage, the proper procedures should be selected and applied.

TABLE 4.5
The Results for Components Not Available in the TOPSIS Method

Component	Score	Unit Cost	Component	Score	Unit Cost	Component	Score	Unit Cost
1	0	37.59456	163	0.002969	1,026	1,943	0.021449	18,759
8	0	222.1698	184	0.002969	858	1,978	0.034508	17,017
9	0	47.73522	193	0.003463	1,112	2,063	0.057140	1,531,832
10	0	1.464041	195	0.003463	1,645.956	2,088	0.035718	18,214
12	0	1206	206	0.003463	726	2,118	0.057322	4,855,416
13	0	327	219	0.003463	728.8557	2,550	0.025268	16,403
15	0	0.915567	221	0.003956	1,017.794	2,606	0.026219	2,360,336
16	0	397	232	0.003956	2,889	2,660	0.026694	1,308,902
19	0	2.1044	274	0.004449	4,495	2,728	0.038627	7,408,593
20	0	7.429627	285	0.004449	918	2,742	0.038627	24,529
24	0	202.632	303	0.004941	969	2,856	0.039646	11,625
25	0.000496	117.35	314	0.004941	1,212.375	2,918	0.030479	33,563
30	0.000496	214	320	0.004941	899	3,059	0.031892	32,090
31	0.000496	1881	380	0.005925	1,893	3,101	0.032362	8,351,062
33	0.000496	143.9316	381	0.005925	1,187	3,109	0.032362	9,717
34	0.000496	406	435	0.006416	1,877	3,160	0.042461	8,025
36	0.000496	107.3312	450	0.006416	1,180.542	3,161	0.061543	100,368
45	0.000496	784	468	0.006907	2,341	3,290	0.034708	115,628
47	0.000496	2,207	508	0.007397	1,804.07	3,420	0.063255	15,857
52	0.000496	430.4264	568	0.008377	1,589	3,450	0.063508	14,344
56	0.000991	419.1864	592	0.028328	16,455	3,510	0.037975	6,719
59	0.000991	512	600	0.008866	2,637	3,764	0.040298	45,740

(Continued)

TABLE 4.5 (*Continued*)
The Results for Components Not Available in the TOPSIS Method

Component	Score	Unit Cost	Component	Score	Unit Cost	Component	Score	Unit Cost
77	0.000991	327	637	0.009355	1,646	3,938	0.042611	21,674
78	0.000991	678.2408	739	0.010331	2,922	4,179	0.045374	27,294
79	0.000991	641	754	0.010331	4,205	4,299	0.047208	176,578
81	0.001486	1,020	763	0.010331	1,724	4,306	0.053867	2,563,771
86	0.001486	609	1,121	0.013251	3,732	4,411	0.070440	127,981
88	0.001486	262.2568	1,141	0.013251	3,377	4,562	0.050404	149,494
89	0.026995	236,947,3	1,192	0.030355	5,948	4,872	0.053128	3,574,578
100	0.001486	584	1,321	0.031037	6,167	5,048	0.055388	2,350,369
103	0.001486	688	1,387	0.016640	3,105	5,073	0.061464	1,031,394
110	0.001981	969	1,415	0.031519	36,675.54	5,332	0.058984	12,040
119	0.001981	6,102	1,454	0.055065	16,507.6	5,355	0.059431	76,891
134	0.002475	452.6001	1,456	0.017123	4,595	5,397	0.059879	44,328
136	0.027068	23,312	1,655	0.055348	11,970	5,449	0.078909	19,495
141	0.002475	652	1,673	0.018087	14,743.07	5,522	0.064333	1,268,945
144	0.002475	507.8689	1,701	0.018569	3,293.592	5,655	0.103221	2,348,774
151	0.027068	6,303,837	1,766	0.033083	43,202	161	0.002969	486
158	0.002969	1,632,057	1,854	0.020490	5,197	1,889	0.020490	446,173

TABLE 4.6
The Results for Components Not Available in the MOORA Method

Component	Score	Unit Cost	Component	Score	Unit Cost
12	0.000663	1,206	2,118	0.005981	4,855,416
31	0.000696	1,881	2,550	0.002385	16,403
47	0.000696	2,207	2,606	0.002452	2,360,336
89	0.002609	236,947.3	2,660	0.002485	1,308,902
119	0.000795	6,102	2,669	0.002518	4,336,173
135	0.002676	4,431	2,728	0.004432	7,408,593
136	0.002676	23,312	2,742	0.004432	24,529
151	0.002676	6,303.837	2,856	0.004531	11,625
158	0.000861	16,320.57	2,918	0.00275	33,563
195	0.000894	1,645.956	3,059	0.002849	32,090
232	0.000928	2,889	3,101	0.002882	8,351,062
274	0.000961	4,495	3,109	0.002882	9,717
380	0.00106	1,893	3,160	0.004796	8,025
435	0.001093	1,877	3,161	0.006643	100,368
468	0.001126	2,341	3,250	0.003015	4,614,087
508	0.00116	1,804.07	3,290	0.003048	115,628
592	0.003106	16,455	3,420	0.006875	15,857
600	0.001259	2,637	3,450	0.006909	14,344
739	0.001358	2,922	3,465	0.003214	6,063,598
754	0.001358	4,205	3,510	0.00328	6,719
1121	0.001557	3,732	3,764	0.003445	45,740
1141	0.001557	3,377	3,829	0.003512	5,939,081
1192	0.003471	5,948	3,838	0.003512	577,455
1321	0.00357	6,167	3,938	0.003611	21,674
1387	0.001789	3,105	4,179	0.00381	27,294
1415	0.003636	36,675.54	4,299	0.003942	176,578
1454	0.005517	16,507.6	4,306	0.00579	2,563,771
1456	0.001822	4,595	4,411	0.007737	127,981
1655	0.005583	11,970	4,562	0.004174	149,494
1673	0.001888	14,743.07	4,673	0.004274	8,193
1701	0.001922	3,293.592	4,872	0.004373	3,574,578
1766	0.003835	43,202	5,048	0.004539	2,350,369
1854	0.002054	5,197	5,332	0.004804	12,040
1889	0.002054	4,461.73	5,355	0.004837	76,891
1943	0.00212	18,759	5,397	0.00487	44,328
1978	0.004001	17,017	5,449	0.008631	19,495
2063	0.005948	153,183.2	5,522	0.005201	1,268,945
2088	0.004133	18,214	5,655	0.01095	2,348,774

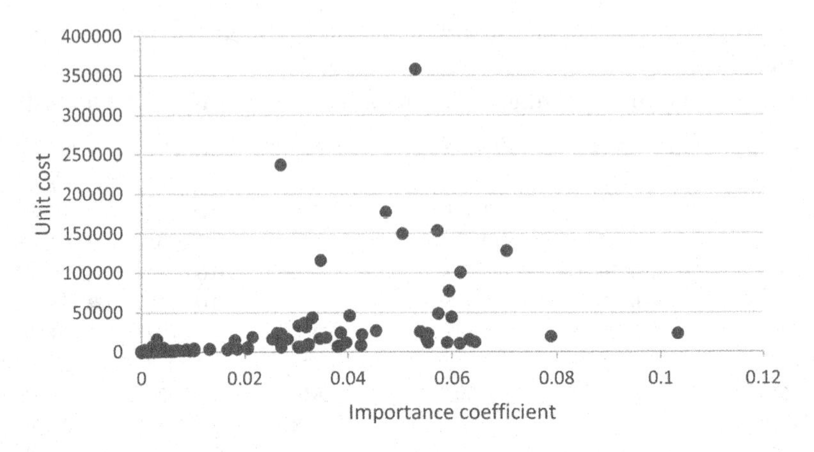

FIGURE 4.2 Unit cost versus importance coefficient for unavailable products in TOPSIS method.

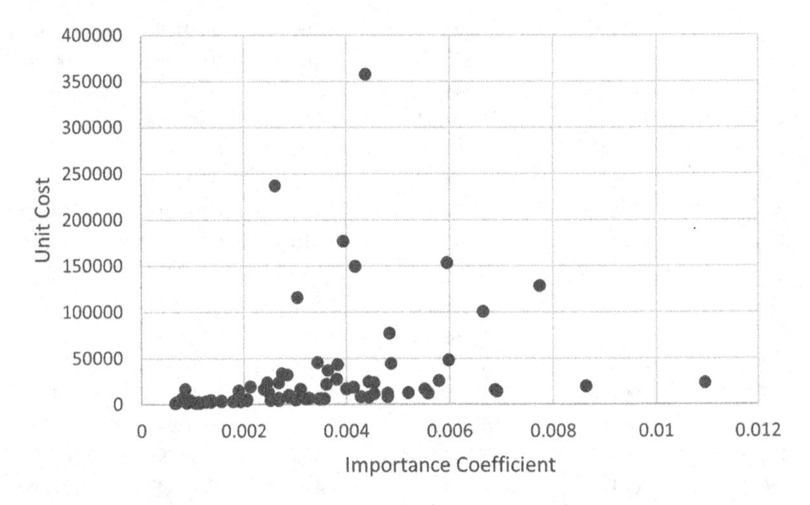

FIGURE 4.3 Unit cost versus importance coefficient for unavailable products in MOORA method.

Therefore, this study proposes using mathematical programming models for spare parts management. These models can optimize the amount and cost of spare parts inventories, simulate various scenarios, and analyze the consequences of these scenarios to aid decision-making. They can also be used for different optimization goals, such as minimizing costs, maximizing customer service levels, or increasing inventory turnover. This study also includes multi-criteria decision-making techniques to reduce uncertainty in the decision-making process. These techniques provide managers with a systematic approach to decision-making and minimize the impact of the human factor by making the decision-making process more objective.

This study has suggested an effective approach for managing obsolescence. Manufacturers of armored vehicles in this sector can use the approach developed in this study to improve decision-making and reduce the risk of shortages of spare parts.

REFERENCES

1. Hu, Q., Chakhar, S., Siraj, S., Labib, A.: Spare parts classification in industrial manufacturing using the dominance-based rough set approach. *European Journal of Operational Research*, 262(3), 1136–1163 (2017).
2. Hu, Q., Boylan, J. E., Chen, H., Labib, A.: OR in spare parts management: A review. *European Journal of Operational Research*, 266(2), 395–414 (2018).
3. Rojo, F. R., Roy, R., Kelly, S.: Obsolescence risk assessment process best practice. *Journal of Physics: Conference Series*, 364(1), 012095 (2012).
4. Braglia, M., Grassi, A., Montanari, R.: Multi-attribute classification method for spare parts inventory management. *Journal of Quality in Maintenance Engineering*, 10(1), 55–65 (2004).
5. Van der Auweraer, S., Boute, R. N., Syntetos, A. A.: Forecasting spare part demand with installed base information: A review. *International Journal of Forecasting*, 35(1), 181–196 (2019).
6. Supçiller, A., Çapraz, O.: AHP-TOPSIS yöntemine dayali tedarikçi seçimi uygulamasi. *Istanbul University Econometrics and Statistics e-Journal*, (13), 1–22 (2011) (In Turkish).
7. Dhakar, T. S., Schmidt, C. P., Miller, D. M.: Base stock level determination for high-cost low demand critical repairable spares. *Computers & Operations Research*, 21(4), 411–420 (1994).
8. Kasap, N., Biçer, İ., Özkaya, B.: Stokastik envanter model kullanılarak iş makinelerinin onarımında kullanılanılan kritik yedek parçalar için envanter yönetim sistemi oluşturulması. *İstanbul Üniversitesi İşletme Fakültesi Dergisi*, 39(2), 310–334 (2010) (In Turkish).
9. Sandborn, P.: Design for obsolescence risk management. *Procedia CIRP*, 11, 15–22 (2013).
10. Ghare, P. M.: A model for an exponentially decaying inventory. *Journal of Industrial Engineering*, 14, 238–243 (1963).
11. Wind, Y., Saaty, T. L.: Marketing applications of the analytic hierarchy process. *Management Science*, 26(7), 641–658 (1980).
12. Lai, Y. J., Liu, T. Y., Hwang, C. L.: Topsis for MODM. *European Journal of Operational Research*, 76(3), 486–500 (1994).
13. Brauers, W. K., Zavadskas, E. K.: The MOORA method and its application to privatization in a transition economy. *Control and Cybernetics*, 35(2), 445–469 (2006).

5 Quality Management Systems

Implementation Based on Opportunity Life Cycle

Pawel Królas and Janne Heilala

5.1 INTRODUCTION

Each of the organization is searching for the opportunity to increase the value to the organization. Opportunities can arise from various factors, including changes in technology, government policies, social trends, and economic conditions. Organizations should constantly monitor their macro and micro environment to identify threats and opportunities.

Threats are external factors that can negatively affect the company, such as new competitors entering the market, changes in regulations, or economic downturns. Threats can also be identified and analyzed, making it possible for the company to take appropriate actions to mitigate their impact.

To capitalize on opportunities and mitigate threats, companies can undertake various actions such as market research to gain insights into customer needs, or invest in new technologies to improve product quality or reduce costs. Additionally, companies can engage in strategic partnerships, expand their geographical reach, or innovate and introduce new products and services to diversify their revenue streams.

Another critical action companies can take is to continuously monitor and evaluate their performance. This allows them to identify areas for improvement and measure the effectiveness of their strategies. By analyzing data and comparing performance against competitors, companies can adjust their strategies and make informed decisions based on current market trends (Królas, 2019).

Therefore, it is essential for companies to identify the life cycle of opportunities in their environment and make appropriate decisions. Short life cycle opportunities may require quick action, while longer life cycle opportunities need more careful consideration and planning. By understanding the life cycle of opportunities, companies can make informed decisions and gain a competitive advantage in their industry (Królas and Włodarkiewicz-Klimek, 2015a).

The establishment of a long-term cycle cooperation between supplier and client has numerous benefits for both parties. The supplier gains stability and predictability of demand, which enables them to plan their production and investments more effectively. This, in turn, leads to better quality of products or services, and often to lower costs

DOI: 10.1201/9781003505327-5

due to economies of scale and scope. The client, on the other hand, gains better access to resources, technology, and expertise of the supplier. They are also able to build a more strategic relationship with the supplier, which enables them to align their objectives, share risks, and achieve mutual benefits. To ensure the success of a cycle cooperation, both parties need to establish clear and transparent communication channels, set up joint performance and quality measures, and continuously monitor and evaluate the cooperation outcomes. Also, both parties need to align their values, culture, and ethical standards to build trust and commitment toward each other. By doing so, they create a sustainable and profitable business model that delivers value to both parties and contributes to the prosperity of the entire supply chain (Królas and Heilala, 2023).

5.2 OPPORTUNITY AND ITS EXPLANATION

The concept of opportunity sparks interest across a range of scientific disciplines, including microeconomics, resource theory, and strategic management (Hunter, 2013). In the field of literature, there exists a diverse array of interpretations and definitions of what constitutes an opportunity (Table 5.1).

TABLE 5.1
Selected Definitions of Opportunity

Source of Definition	Definition of Opportunity
Green	Refers to the potential creation of a new venture. This applies to both business and non-profit activities, ventures that provide value to both customers and owners of the organization
Hunter	It is related to the individual perception of reality, the ability to perceive favorable conditions which, through the use of specific concepts, can be presented in the form of ideas
Trzcieliński	A situation that supports the agency of action in attaining the intended outcome or desired impact, which either exists within the surroundings of the entity or is a hypothetical state of the characteristics of this environment
Sull	It results from a new mix of innovative resources, also taking into account technologies, processes, supply chain, business model, maintaining or disrupting the current state by extending practices and/or "breaking with the past" to meet new customer needs
Krupski	It is an event (e.g., collapse of a competitor) or a combination of various circumstances (e.g., creation of a market niche) of an economic nature (or with economic consequences), creating opportunities to achieve additional benefits, material, and/or intangible values
Eckhardt, Shane	It is a situation in which new products, services, materials, markets, and organizational methods can be implemented by formulating new meanings for them
Koen and others	It is a business and/or technological space (gap) between the current situation and the vision of the future recognized by an enterprise or an individual. By using competitive advantage, it is possible to eliminate threats and solve the problem

Source: Own study based on Green (2015), Hunter (2013), Trzcieliński (2011), Sull (2009), Krupski (2005), Eckhardt and Shane (2003), Koen (2001).

The presented definitions show different approaches to the meaning of the opportunity. Each of the definitions shows the benefit to the organization. The occurrence of the opportunity itself is objective, but "capturing" and utilizing it is subjective. The subjectivity of the opportunity depends on the place where it occurs (Figure 5.1).

Opportunities that arise within the organization have a high level of subjectivity; in a competitive environment, it is moderate; and in the macro-environment, it is low.

Long time of the duration of the opportunity creates the conditions for more players to take advantage of it. Such an opportunity is characterized by a very low level of subjectivity. The shorter duration of the opportunity allows it to be used by only a few companies. Therefore, the possibility of realizing the opportunity depends on the subjective assessment of the organization (high level of subjectivity) (Karpacz, 2010). On otherwise, Skat-Rørdam distinguishes between opportunities of a tactical and strategic nature (Figure 5.2).

Opportunities of strategic importance are related to a radical change in the way the company participates in the competitive game. This may also provide a significant portion of the company's revenue in the long-term perspective. The remaining identified opportunities have tactical significance.

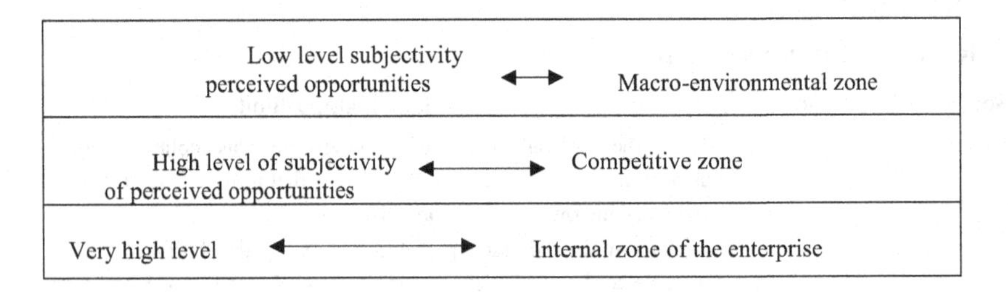

FIGURE 5.1 The level of subjectivity of opportunity perception. Adapted from Karpacz (2010).

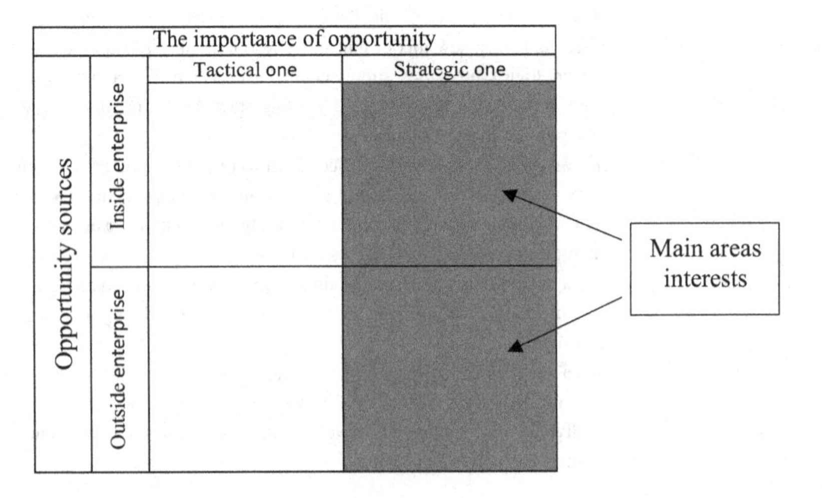

FIGURE 5.2 Opportunity space. Adapted from Skat-Rørdam (2001).

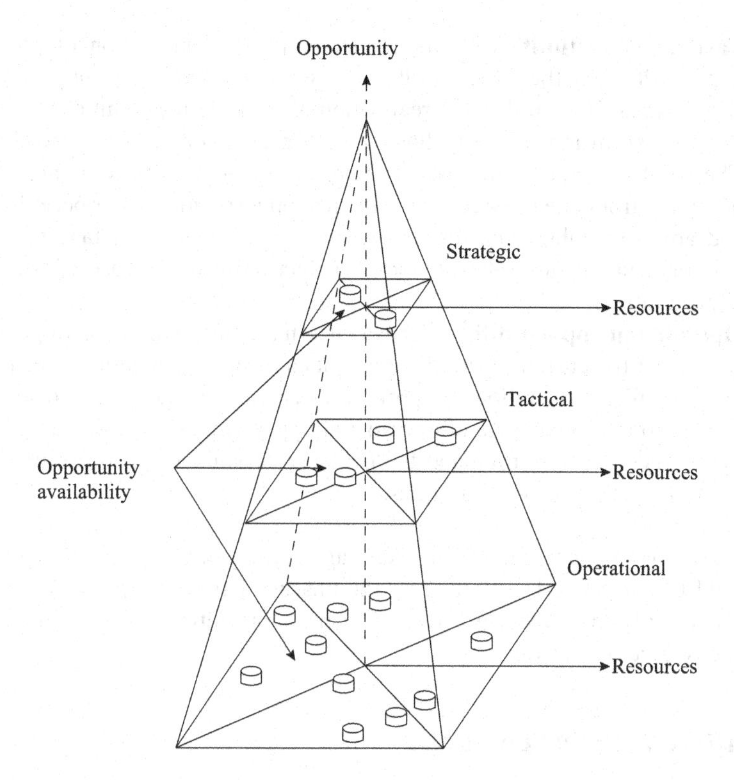

FIGURE 5.3 The nature of the opportunity and the availability of resources. Own study based on Królas (2015b).

Opportunities can also be categorized by the levels of management within the organization, particularly from the perspective of the hierarchy level and the necessary resources required to pursue the opportunity. Thus, depending on the level of management, three types of opportunities can be distinguished (Figure 5.3).

- **Strategic opportunities** – They are related to a significant change in the functioning of the enterprise; from a market perspective, taking advantage of these opportunities means strengthening the company's position on the market, ensuring faster growth of the company than the average in the industry, strengthening the organization's reputation toward customers, dramatically improving the technology of manufactured products/services; in financial terms, this means achieving higher revenues/profits as well as lower total costs, improving financial liquidity; in practice, this may concern the acquisition of an enterprise, diversification of activities, and entry into new markets. This constitutes a qualitative leap in the current activity. Strategic opportunities occur very rarely; taking advantage of opportunities is related to knowledge of the market and customer preferences, as well as extensive relationship capital of the organization's managers; the time horizon covered by this type of opportunity covers from several to a dozen or so years; a financial opportunity requires large financial outlays, and their return is spread over time. When the project fails, there may be great difficulties in ensuring business continuity.

- **Tactical opportunities** – They concern the implementation of projects that result from the decomposition of strategic goals into subject plans (Kałkowska, 2010, p. 12); the realization of tactical opportunities is related to investing in innovative technology and establishing a new partnership. The effects related to the use of opportunities usually have a short time horizon – from one to three years. Tactical opportunities occur periodically and are predictable. The resources necessary to take advantage of tactical opportunities are relatively smaller than those in the case of strategic opportunities.
- **Operational opportunities** – They concern the implementation of projects that affect the current operational activities of the organization. The number of available operational opportunities is relatively large. The resources needed to take advantage of operational opportunities are relatively small; they largely concern time and the necessary organization of resources for preparing and using opportunities.

The duration of an opportunity's life cycle can vary depending on the type of opportunity and the industry. Understanding the phases of the life cycle can help businesses make informed decisions about which opportunities to pursue and how to allocate resources effectively.

5.3 LIFE CYCLE OPPORTUNITY

The interconnection of the opportunity's life cycle with the industry and external factors has an impact on the organization. Factors such as the level of assurance and longevity of the beneficial circumstance determine the duration of the opportunity's life cycle (Figure 5.4; Trzcieliński, 2011).

Life cycle opportunities can be divided into distinct phases (Trzcieliński, 2011):

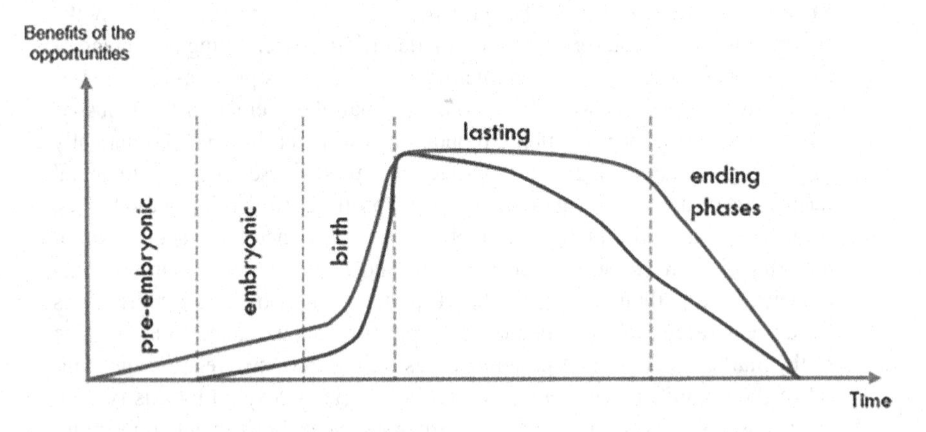

FIGURE 5.4 Life cycle opportunity. Adapted from Trzcieliński (2011).

- **Pre-embryonic** – At this stage, there are announcements of the events in the future, often with a long-time horizon, that could be considered favorable for the organization. In the pre-embryonic phase, the risk is related to the need to set up the necessary resources before the opportunity arises. The time of the opportunity birth phase at which the company can consider the emerging situation as complete and favorable is unknown.
- **Embryonic** – This phase focuses on the arising of events that can be considered as leading to the emergence of an opportunity. The risk concerns resources and the final parameters of the opportunity. Risk management comes down to securing the funds that are necessary to take advantage of opportunities.
- **Duration** – The risk associated with the existence phase concerns the ability to take advantage of the opportunity (having the necessary potential).
- **Decline** – In this phase, changes could occur in the relationships due to factors such as situational impact, reality of the goal, and availability of resources, which occurred during the opportunity phase. The risk associated with the decline phase includes changes in the relationships between situational impact, reality of the goal, and adequacy of resources. Risk management includes the elimination of connections between partners (other people's resources), the effects of using opportunities, and the possibility of reconfiguring resources in order to use another opportunity (Trzcieliński, 2011, p. 63).

When a product or service is in the beginning stages of its life cycle, there is often a unique opportunity for suppliers and clients to form a cooperative relationship. This collaboration can be beneficial for both parties involved (Figure 5.5).

In Figure 5.5, the collaboration between the supplier and customer is depicted. The supplier seeks to derive value from the cooperation, such as acquiring new clients, retaining existing clients, and establishing a strong market position, while also generating profits for the organization.

Each partnership has a specific start and end date. Based on Figure 5.5, the value of the cooperation may be relatively low, but it increases over time before declining toward the end of the cooperation. The duration of the collaboration is determined by the agreement between the supplier and the client.

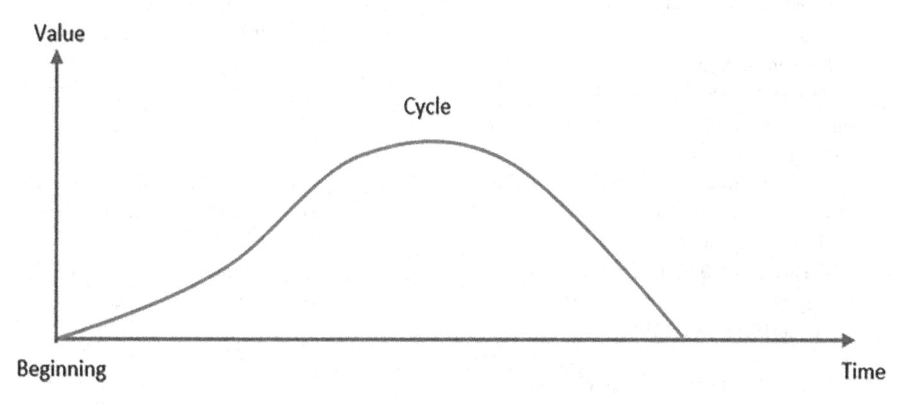

FIGURE 5.5 Cooperation between supplier and customer. Sources: Based on own study.

There are different forms of cooperation between suppliers and customers depending on the sector of industry. Quality Management Systems (QMS) create three key areas that typically collaborate with one another:

- **Quality management unit** – Client organization where QMS is implemented.
- **Implementation partner** – A company that implements the QMS in client organization.
- **Certifying agency** – The Compliance Certification Authority.

Over time, there have been changes to the number and types of quality management systems and certificates.

5.4 QUALITY MANAGEMENT SYSTEMS

In the early 1990s, Polish organizations began implementing ISO's quality management system (QMS). Initially, the larger production organizations were the main players in implementing the ISO standards. By adopting the ISO standards, enterprises were able to improve their processes, focus on objectives, improve the skill set of their workforce, and establish cycles of action mandated by ISO specifications (Królas, 2020).

In recent years, an increasing number of small- and medium-sized organizations has embraced ISO systems. These organizations have not only integrated ISO 9001, but also expanded their adoption to include ISO 14001 for Environmental Management Systems and ISO 45001 for Occupational Health and Safety Management Systems. The issuance of certificates to Polish organizations has experienced noticeable fluctuations, as evident in Table 5.2.

TABLE 5.2
Number of Certificates in Poland between 2017 and 2021

ISO	2017	2018	2019	2020	2021	2021/2020 (%)
Quality Management System ISO 9001:2015	11,846	11,294	11,460	10,219	10,512	+2
Environmental Management Systems ISO 14001:2015	2,885	2,921	3,466	2,748	2,831	+3
Information Security Management System ISO 27001:2017	705	700	652	710	876	+23
Occupational Health and Safety Management Systems ISO 45001:2018	–	83	257	1,141	1,646	+44
Energy Management Systems ISO 50001:2018	173	139	141	178	250	+40

Source: Own study based on www.iso.org.

One of the most popular standards implemented in polish organization is ISO 9001:2015. It is often a requirement for organizations to possess ISO 9001 in order to qualify for public tenders. In 2021, a total of 10,512 certificates were issued by certification bodies, representing a slow increase of 2% compared to the previous year. Over the past 5 years, the number of issued certificates has remained consistent, ranging from 10,000 to 12,000.

A similar situation can be observed in Environmental Management Systems with ISO 14001. The number of issued certificates has steadily increased by 3% each year since 2020, ranging from 2,885 to 3,466 certificates between 2017 and 2021.

Both ISO 9001 and ISO 14001 are important standards for organizations in Poland. While the number of issued certificates has been relatively consistent for ISO 9001, there has been a steady increase in the number of certificates issued for ISO 14001.

Information Security Management System ISO 27001:2017 (current ISO 27001:2022) has experienced a significant increase of 23% in the number of its certificates from 2020 to 2021. Although the quantity of certificates is relatively small when compared to ISO 9001 or ISO 14001, there has been a marked trend toward greater significance being placed on safety in relation to both software and hardware within business operations. This suggests that companies are increasingly aware of the importance of safeguarding information and protecting against cybersecurity threats, leading to greater uptake of this particular standard.

The situation for other management systems, such as Occupational Health and Safety Management Systems ISO 45001:2018 and Energy Management System ISO 50001:2018, shares similarities. These standards were established in 2018 and organizations were given three years to transition to the new versions, during which the number of certificates steadily rose. In 2021, there was a significant 40% increase in the number of certificates compared to the previous year, suggesting that companies are recognizing the importance of these management systems in maintaining health and safety standards, as well as reducing energy consumption and waste in a sustainable manner.

Table 5.1 in this article shows the number of certificates for different ISO standards issued by certification bodies. There are many organizations that introduce various ISO standards to ensure that the requirements of ISO are being met. However, due to the high cost associated with joining certification body audits, some organizations choose not to participate in the audits.

Regular cycle activities within QMS necessitate annual maintenance of the certificate. These activities encompass audits and management reviews, which can be undertaken internally by the organization or outsourced to an external advisory unit. Regardless, the required competencies need to be present to ensure that these actions are carried out effectively. Small- and medium-sized businesses often use external advisor companies to assist with this process.

On other side, large organizations typically have more financial resources compared to smaller organizations. This allows them to invest in training and development programs to ensure that their employees possess the necessary competencies for dealing with QMS standards. Therefore, large organizations can often carry out cycle actions within QMS by themselves. The importance of maintaining QMS standards is emphasized in the article, and cycle actions are necessary to ensure that the organization is adhering to the requirements of the ISO standards.

5.5 CASE STUDY

The study took place in the advisory firm, SIMPTEST Poznan, where they introduced QMS in Alpha company (the authors of the paper did not receive consent to publish the company name). SIMPTEST Poznan offers services related to environmental management, health and safety management, information security management, and process optimization. The company specializes in providing customized solutions to meet the specific needs of its clients, and its team of experts works closely with organizations to ensure that their QMS is effectively implemented and maintained. SIMPTEST Poznan is committed to providing high-quality services that meet the highest industry standards and to helping its clients achieve their business objectives through effective quality management practices. SIMPTEST was founded in the year 1983.

Alpha company is a company that has expertise in the upkeep, installation, and maintenance of wind turbines, as well as blade repairs. Alpha has been involved in numerous installations globally, encompassing more than 1,000 wind turbines throughout Europe, Africa, and Asia. The study sought to exhibit how SIMPTEST Poznan implemented QMS within Alpha, as a means of capitalizing on the life cycle potential as depicted in Figure 5.6.

From 2020 to 2022, a comprehensive study was carried out to introduce the QMS, namely ISO 9001, ISO 14001, and ISO 45001, within the Alpha organization. The project manager responsible for overseeing the development and integration of QMS in the Alpha organization was also one of the co-authors mentioned in Figure 5.7.

Each of the introduced QMS in client organization consists of different phases depending on the needs of the organization.

The case study showed the implementation consisting of audit, training courses, preparations of the documentations, consultations, and management reviews (Figure 5.7).

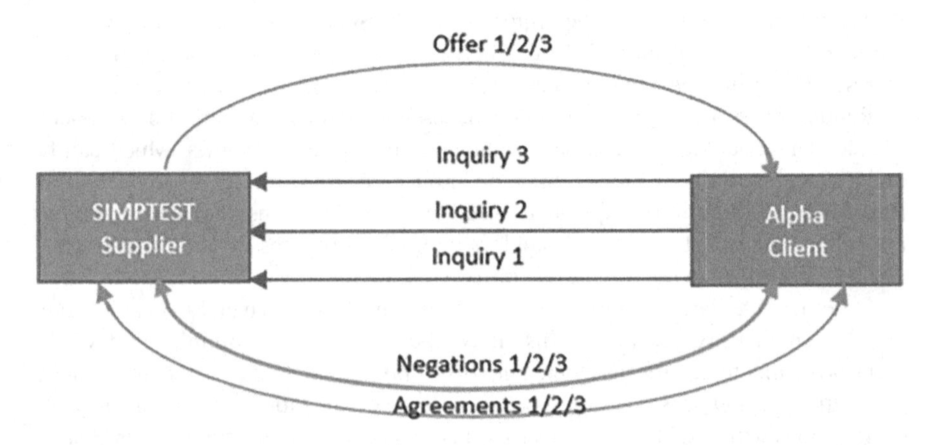

FIGURE 5.6 Cooperation between the companies. Source: Own study.

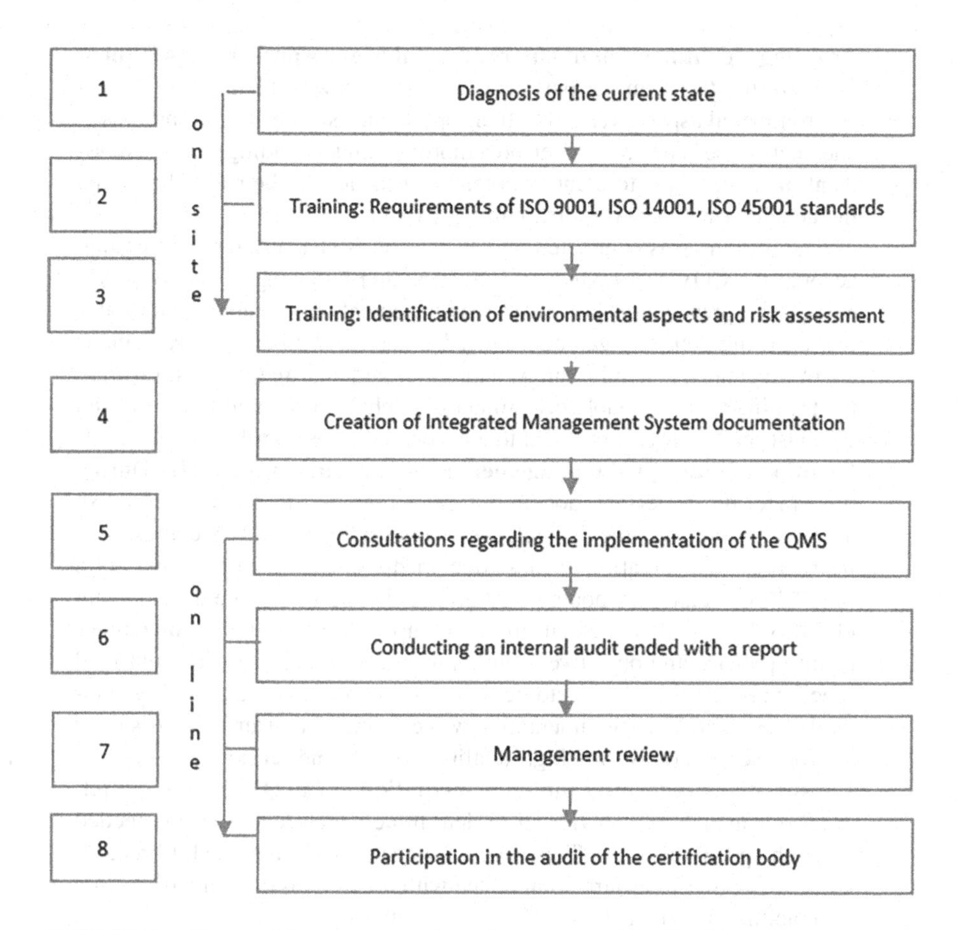

FIGURE 5.7 Phases of development and implementation of Quality Management Systems. Source: Own study.

The implementation consisted of eight phases:

- **Diagnosis of the current state (1)** – The consultant firm has gained a comprehensive understanding of the operations of Enterprise Alpha, including its strategies, organizational framework, protocols, and guidelines. An interview was conducted with the members of the management board, along with selected managers and experts, at Alpha's headquarters. The final element of this stage involved generating a report that demonstrated the extent to which ISO 9001, ISO 14001, and ISO 45001 standards had been met.
- **Training: Requirements of ISO 9001, ISO 14001, ISO 45001 standards (2)** – During this phase, representatives from the advisor company underwent a training course that focused on the requirements of the relevant standards. The course covered topics such as the implementation of QMS, regulatory compliance, risk management, and other relevant topics. The training provided the representatives with an understanding of the standards, their implications, and the steps required to meet the requirements. The goal of this training was to ensure that the representatives were equipped with the knowledge and skills needed to effectively assist their clients in complying with the standards.

- **Training: Identification of environmental aspects and risk assessment (3)** – During the training, participants learned how to identify and assess environmental aspects related to their operations, such as emissions, waste, and energy use. They also received a thorough understanding of risk assessment, including how to identify potential risks, assess their likelihood and impact, and establish controls to mitigate them.

  The training was conducted on-site at Alpha's headquarters, taking into account the COVID-19 pandemic situation and following all safety protocols. The representatives of Alpha's management participated in the training, ensuring that they gained knowledge and skills to effectively manage environmental risks and comply with environmental regulations. Overall, the training was a valuable investment in Alpha's sustainability efforts and demonstrated their commitment to environmental responsibility.
- **Creation of integrated management system documentation (4)** –During this stage, the necessary documents were prepared in order to establish a QMS. These documents included a quality manual, procedures, and instructions. The creation of these documents was done with the help of a SIMPTEST Poznan company, who provided expertise in the development of QMS. The quality manual provided an overview of the organization's quality policies and objectives, while procedures and instructions outlined specific processes and tasks to be followed in order to ensure quality standards were met. This documentation was essential for ensuring consistency in processes and delivering high-quality products and services.
- **Consultations regarding the implementation of the QMS (5)** – The relevant documents were reviewed to determine which records were needed for each specific process. This suggested a systematic approach to record-keeping in order to ensure compliance with regulations or standards, while also facilitating efficient and effective operations.
- **Conducting an internal audit ended with a report (6)** – The report of the internal audit provided a comprehensive analysis of the level of compliance with the relevant standards. The audit team identified areas of strengths within the organization that were consistent with the standards. Additionally, they highlighted areas that had potential for improvement to meet the requirements set by the standard. The report from the audit also identified non-conformities present within the organization that needed corrective action to bring them into compliance with the standards. Non-conformities can be any deviation from the established requirements, and they were identified and documented using non-conformity cards which were discussed with the responsible departments or individuals. Corrective actions were taken to address the non-conformities and ensure compliance with the standards.
- **Management review (7)** – The management review evaluated aspects typical of the relevant standards pertaining to quality, environment, and health and safety.
- **Participation in the audit of the certification body (8)** – During the external certification body's audit, Alpha received support from representatives of the consulting company which led to the company obtaining a certificate of compliance with the applicable standards.

The COVID-19 pandemic has required many organizations to adapt and conduct their operations remotely, including audits. Therefore, phases 5–8 of the audit were conducted online to ensure the safety of all parties involved.

The internal audit and management review was carried out by SIMPTEST Poznań according to the agreed terms and schedule. The audit report was submitted to the Alpha company, highlighting areas of improvement and making recommendations for enhancing the company's processes and procedures. The management review was also conducted, with the Alpha company's management team discussing and reviewing the audit results, as well as planning actions to address the recommendations. The implementation of these actions was monitored by SIMPTEST Poznań, with periodic follow-up audits and reviews conducted to ensure continuous improvement of the company's system. Overall, the procedure proved to be effective in helping the Alpha company maintain a high level of quality and efficiency in its operations.

Figure 5.8 displays the life cycle and use of the opportunity based on the presented case study, which lasted for a period of 2 years (2021–2022).

Cycles II and III were initiated by an inquiry from Alpha organization to conduct a supervision audit and management review of their QMS based on ISO 9001, ISO 14001, and ISO 45001. The advisor company prepared an offer based on their experience and knowledge of the client's organization and the market situation. Negotiations took place before the agreement was signed. The scope of work included two stages: supervision audit and management review. The timing of the work was set for one month before the external audit by the certification body. Cycles II and III took place in October/November 2021 and October/November 2022, respectively. The life cycle and use of opportunity for each cycle was approximately 2 months.

Based on presented case study, it is possible to create further cooperation between client and supplier based on a win-win situation.

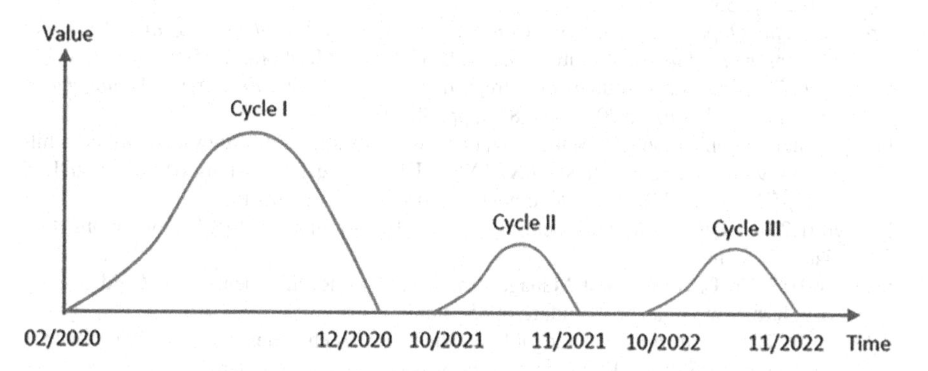

FIGURE 5.8 Life cycle opportunity based on implementation and review of QMS. Sources: Own study.

5.6 SUMMARY

Expanding upon the notion of opportunity, it can be described as a favorable situation or circumstance that aligns with an individual's or organization's objectives and allows for the possibility of achieving their intended goals. The utilization and development of these opportunities are contingent upon a variety of factors, including the resources and timing available to the organization, as well as the manner in which the opportunity was created.

Through effectively capitalizing on these opportunities, businesses can establish cooperative relationships between suppliers and clients. Depending on the quality of these relationships and their appeal, they can either lead to self-contained projects or to a cycle of continuous collaboration. A prime example of how this process operates can be seen through the implementation of QMS.

The implementation of QMS can be divided into multiple phases, varying in length and complexity depending on the goals and requirements of the organization. A case study focusing on the implementation of QMS exemplifies this concept by highlighting a process consisting of seven key phases: the diagnosis of the current state, two training courses, the creation of integrated management system documentation, consultations regarding QMS implementation, conducting an internal audit, management review, and participation in the audit of the certification body.

By establishing these cooperative relationships through the effective implementation of QMS, advisor companies (suppliers) can successfully create new partnerships with clients such as the Alpha organization mentioned in the case study. Furthermore, such relationships can yield favorable and optimistic prospects for future partnerships.

Overall, this approach of identifying and implementing life cycle opportunities presents a valuable viewpoint for businesses across different industries, inspiring them to establish business cooperation based on this model.

REFERENCES

Eckhardt J., Shane S., Opportunities and entrepreneurship, *Journal of Management* 2003, vol. 29, p. 336.

Green J., *The Opportunity Analysis Canvas. A New Tool for Identifying and Analyzing Entrepreneurial Ideas,* Venture Artisans, LLC, Upper Marlboro, 2015, p. 1.

Hunter M., Typologies and sources of entrepreneurial opportunity, *Economics, Management, and Financial Markets* 2013, vol. 8(3), pp. 58–59.

International Organization for Standardization, www.iso.org, https://www.iso.org/committee/54998.html?t=dnBm2j_sAhhXB1XFYcHFdz9kmqJlQH9v-kmsAvQsa1mCgDLySpPIcp5ZMcOBvnI3&view=documents#section-isodocuments-top

ISO 9001:2015, Quality Management Systems - Requirements. https://www.iso.org/standard/62085.html

ISO 14001:2015, Environmental Management Systems - Requirements with Guidance for Use, https://www.iso.org/standard/60857.html

ISO 27001:2022, Information Technology. Security Techniques. Information Security Management Systems. Requirements. https://www.iso.org/standard/27001

ISO 45001:2018, Occupational Health and Safety Management Systems. Requirements with Guidance for Use. https://www.iso.org/standard/63787.html

ISO 50001:2018, Energy Management Systems - Requirements with Guidance for Use. https://www.iso.org/iso-50001-energy-management.html

Kałkowska J., *Zarządzanie strategiczne. Metody analizy strategicznej z przykładami*, Wydawnictwo Politechniki Poznańskiej, Poznań, 2010, p. 12.

Karpacz J., Determinanty dostrzegania i wykorzystania okazji przez przedsiębiorców. In: A. Stabryła (ed.), *Koncepcja zarządzania współczesnym przedsiębiorstwem*, seria "Encyklopedia Zarządzania", Mfiles.pl, Kraków, 2010, p. 47.

Koen P., Providing clarity and a common language to the fuzzy front end, *Research Technology Management*, 2001, vol. 44, pp. 46–55.

Królas P., Identyfikacja ryzyka związanego z krótkim cyklem życia okazji - studium przypadku. *Zeszyty Naukowe Politechniki Poznańskiej "Organizacja i Zarządzanie", Wyd* 2019, vol. 80, p. 140.

Królas P., Human aspect and risk in quality management systems. In: S. Trzcieliński and W. Karwowski (eds.) *Advances in the Ergonomics in Manufacturing: Managing the Enterprise of the Future. AHFE (2020) International Conference. AHFE Open Access*, vol. 12, AHFE International, Honolulu, HI, 2020. https://openaccess.cms-conferences.org/publications/book/978-1-4951-2103-6/article/978-1-4951-2103-6_6.

Królas P., Heilala J., Life cycle opportunity based on implementation of quality management systems. In: W. Karwowski and S. Trzcieliński (eds.) *Human Aspects of Advanced Manufacturing. AHFE (2023) International Conference. AHFE Open Access*, vol. 80, AHFE International, Honolulu, HI, 2023, pp. 57–66. https://openaccess.cms-conferences.org/publications/book/978-1-958651-56-8/article/978-1-958651-56-8_5

Królas P., Włodarkiewicz-Klimek H., Okazja a ryzko we współczesnym przedsiębiorstwie, w: Kształtowanie zwinności przedsiębiorstw: monografia. In: S. Trzcieliński (ed.) *Taduesz Zaborowski (WIZ) – Poznań*, IBEN, Gorzów Wlkp., Polska, 2015a, pp. 95–103.

Królas P., Włodarkiewicz-Klimek H., Okazje - źródła, rodzaje i ryzyko, w: Kształtowanie zwinności przedsiębiorstw: Monografia. In: S. Trzcieliński (ed.) *Taduesz Zaborowski (WIZ) – Poznań*, IBEN, Gorzów Wlkp., Polska, 2015b, pp. 94–105.

Krupski R. (ed.), *Zarządzanie przedsiębiorstwem w turbulentnym otoczeniu*, Polskie Wydawnictwo Ekonomiczne, Warszawa, 2005, p. 49.

Skat-Rørdam P., *Zmiany decyzji strategicznych. Wykorzystanie okazji rynkowych do rozwoju przedsiębiorstwa*, Wydawnictwo Naukowe PWN, Warszawa, 2001, pp. 149–150.

Sull D., *The Upside of Turbulence. Seizing Opportunity in an Uncertain World*, Harper Collins Publishers, New York, 2009, pp. 20–21.

Trzcieliński S., *Zwinne przedsiębiorstwo,* Wydawnictwo Politechniki Poznańskiej, Poznań, 2011, pp. 43–60.

6 Speed Up Medical Device Innovation
Combining Agile, User Experience, and Human Factors Methodologies

Adrián Morales-Casas, Amparo López-Vicente, Lorenzo Solano-García, and José Laparra-Hernánez

6.1 INTRODUCTION

Recent analyses underscore the widespread adoption of Lean, Six Sigma, and Lean Six Sigma (LSS) methodologies within the MedTech and Pharma industry, with nearly 95% of organisations benefiting from some form of these frameworks (McGrane et al. 2022; Rathi et al. 2022). The integration of management methodologies and engineering approaches has proven vital for achieving clear objectives in the MedTech field. While various methodologies, including Lean (Anderson et al. n.d.; Slattery et al. 2022), Lean Six Sigma (Byrne et al. 2021), Agile (Gerber et al. 2019; Martens et al. 2022), Stage-Gate (Pietzsch et al. 2009), and hybrids (Cooper and Sommer 2016), have been employed in Medical Device Development (MDD) and Medical Device Implementation (MDI) project management, their effectiveness and adaptability depend on the unique context of each project.

MedTech projects, characterised by short product life cycles, demand innovative solutions to navigate the complexities of regulatory compliance, design improvements, and time-sensitive launches. Existing project management methods, such as Stage-Gate and concurrent engineering, provide operational frameworks, but may face challenges in addressing the dynamic and iterative nature of medical device development. This work aims to explore and present a case study-oriented research approach that aligns with Agile principles and incorporates Human Factors Engineering (HFE) for enhanced usability.

In contrast to the traditional Stage-Gate model, Agile project management, recognised for its adaptability and iterative cycles, emerges as a compelling alternative for overcoming limitations in MDD. This study shares insights from a 16-month project wherein an Agile approach, complemented by HFE, facilitated the development of a non-contact vital signs measuring system. The scope extended from concept

DOI: 10.1201/9781003505327-6

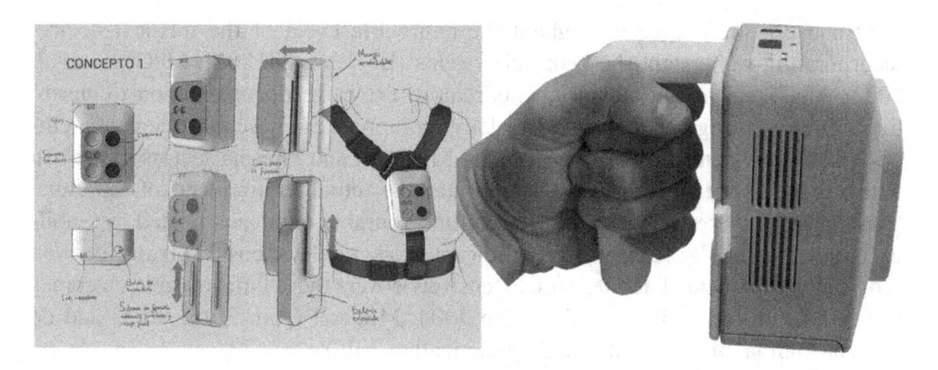

FIGURE 6.1 From concept to alfa non-contact vital signs measuring prototype.

initiation to a TRL 7 stage, presenting an innovative solution capable of measuring vital signs through video images at a distance of 2 m.

The project was executed in collaboration with biosignals solution company Plux, which worked on developing the APP interface, and was promoted by the 061 Health Emergencies Centre of the Andalusian Health Service through the "Promotion of Innovation from Demand" programme. The objective was to create non-contact vital signs measuring system reaching TRL 7, advancing from the conceptual phase to an alpha prototype, as illustrated in Figure 6.1. The prototype had to measure crucial metrics, encompassing heart rate, respiratory rate, oxygen saturation, temperature, and blood pressure, which were captured without physical contact, maintaining a distance of 2 m. Also, to demonstrate its functionality, the prototype was tested in a relevant environment to reach the demanded TRL. The system additionally comprised a locally installed application on a dashboard, overseeing and collecting data from the non-contact measuring device.

6.2 BACKGROUND

The MedTech and Pharma sectors stand out as highly innovation-driven industries, where products typically endure for only 18–24 months before a newer, superior version emerges, as reported by MedTech Europe's Facts and Figures 2022 (MedTech Europe n.d.). Despite the relatively short lifespan of these products, the persistent upsurge in research and development within both industry and academia reflects a robust commitment among all stakeholders in the sector. However, the introduction of novel products brings forth significant challenges. Modifications and advancements in design or manufacturing often demand fresh validation and resubmission to regulatory bodies for thorough review (McGrane et al. 2022; Marešová et al. 2020; Huusko et al. 2023).

Furthermore, managerial approaches rooted in concurrent engineering and the widely embraced Stage-Gate system prioritise tasks and timelines over reasons and methods, as elaborated in the "Design Control Guidance For Medical Device Manufacturers" by FDA (FDA n.d.), along with insights from Pietzsch et al. (2009) and Slattery et al. (2022). Lastly, the adoption of project management strategies or enhancements in MedTech tends to lag behind less-regulated industries. Regardless of whether the project is an MDD or an MDI, the regulatory framework remains rigid and comprehensive throughout the entire life cycle of the medical device, incorporating well-established control systems (Boylan et al. 2021; McGrane et al. 2022). MedTech project advancements require extensive documentation to ensure traceability of design process and legislative compliance. These management methods and unique constraints inherent in the MedTech and Pharma sectors may clash with a project's optimal time-cost balance and the actual requirements of end users. Consequently, there is an elevated risk of unfavourable outcomes related to usability issues, potentially resulting in significant project delays, cost overruns, or even project abandonment. The literature extensively covers these unfavourable outcomes (Lin et al. 2001; Fairbanks and Caplan 2004; Mitchell et al. 2015; Roma and de Vilhena Garcia 2020; World Health Organization 2010).

Fortunately, both industry and numerous academic research studies are exploring ways to expedite the journey of new products from conception to market with fewer errors, leveraging continuous improvement management and engineering methodologies (Slattery et al. 2022). However, a scarcity of case study-focused research that includes detailed process mapping and a comprehensive list of techniques used in implementing solutions in each case persists.

The objective of this study is to present product development research centred on case studies, with a focus on process mapping and the methodologies employed in solution implementation. This will be substantiated by a real case that necessitates simultaneous hardware and software development.

6.3 REVIEW OF MANAGEMENT METHODOLOGIES IN THE MEDTECH

Multiple strategies have been devised and refined to empower managers and developers with the requisite cognitive frameworks, analytical tools, and procedural workflows for discerning optimal solutions. This section unveils contemporary management strategies by delineating their features and practical application within the MDD process. All extant management approaches share a common goal of augmenting quality, meeting customer expectations, abbreviating delivery timelines, curbing costs, and ensuring regulatory compliance to varying extents. Various techniques, including the "Seven Basic Quality Tools" and "Advanced Quality Tools" (Neagoe and Klein 2010), alongside TQM, concurrent engineering, and activity-based costing methods (Sloan 1996), are employed to achieve these objectives. These methodologies furnish managers with diverse perspectives, analytical instruments, and transformation tools (Van Der Peijl 2012). Nevertheless, irrespective of the chosen management approach, one of the initial steps involves establishing technical specifications and user requirements, often constituting the most challenging phase in the entire product development process.

6.3.1 Get the Requirements of the Users Right

The essence of effective problem-solving lies in a range of diverse management strategies. Before delving into these strategies, it is crucial to emphasise the initial step: identifying and understanding problems and needs. As highlighted in the introductory section, the significance of this initial phase is heightened in the MedTech industry. In such a dynamic sector, choosing inappropriate requirements can result in adverse outcomes, customer dissatisfaction, or a regression from an advanced stage back to a complex design or conceptual phase due to the absence of vital features or value. Recognising the pivotal role of this preliminary stage, along with the tools and guidelines to translate requirements into the product, is essential for establishing the foundation for an in-depth exploration of key management strategies and determining the design outcomes.

Ensuring coherence in the requirements that a product must meet involves adhering to five fundamental principles. These requirements-defining principles encompass (1) a deep understanding of the knowledge domain, (2) identification of key stakeholders, (3) analysis of stakeholders' traits and behaviour, (4) delineation of the process to extract requirements from stakeholders using one or more techniques, and (5) gathering requirements from stakeholders or end users (Salleh and Nohuddin 2019). Illustratively depicted in Figure 6.2, these five principles can be encapsulated and understood through the following stages: an initial learning phase, facilitated by the first three principles; an application stage, where principles four and five come into play; and a subsequent validation phase for assessing outcomes. This conventional approach to problem identification and resolution is well-established and widely implemented (Lumsdaine and Lumsdaine 1994).

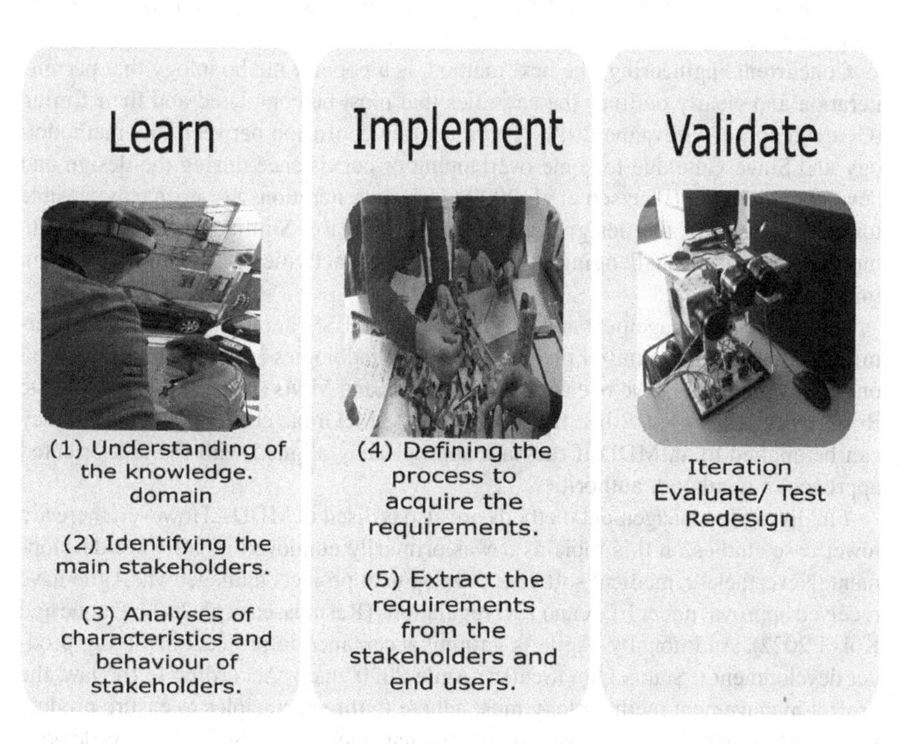

FIGURE 6.2 Stages of requirement acquisition.

The requirements acquisition techniques (RAT) employed in navigating the learning and implementation stages can be classified into two categories: methodologies and design strategies. The primary techniques used in the process of defining requirements encompass interviews, discussions, focus groups, surveys, observations, requirements workshops, and prototyping, among others (Sun et al. 2019). Parallel to the methodologies, RAT design strategies serve as valuable instruments for translating customers' needs into measurable product features, albeit from a design perspective. Noteworthy design strategies include Design for Reliability (DfR), Design for Quality (DfQ), Design for Validation (DfV), and Design for Usability (DfU) (Slattery et al. 2022; Saidi et al. 2019).

6.3.2 Choosing the Proper Problem-Solving Structure

While the stages of learning, executing, and validating remain constant, the degree of latitude to incorporate iterations or checkpoints for product enhancement throughout the process varies depending on the management methodology employed. We will discuss the four main management methodologies in this section. Stage-Gate, the first and most widely embraced management technique, functions as both a theoretical and practical model for transitioning a new product from its conception to its launch (Pietzsch et al. 2009; Slattery et al. 2022). This approach does not account for iterations or the repetition of actions from prior phases. However, it has been successfully deployed and has facilitated regulatory compliance for products that are not overly complex.

Concurrent engineering, the next method, is a popular methodology that permits iteration and clearly outlines the activities that must be completed and their timing (Goldenberg and Gravagna 2018). There is often confusion between this methodology and Stage-Gate due to some overlapping or coexistence during the design and development phase (Pietzsch et al. 2009). However, iterations are primarily feasible towards the end of the design and development phase. Similar to the Stage-Gate methodology, it is challenging to repeat previous activities once they have been implemented.

Thirdly, lean management and its hybrid variant, LSS, are methodologies that permit iterations from the initial phases. These methodologies are consistently focused on improvement and have been applied in MDDs and MDIs (Anderson et al. n.d.; De Rossi 2012; Khan et al. 2013; Byrne et al. 2021; McGrane et al. 2022). Hence, they can be applied to an MDD if they are appropriately aligned with the risk-oriented approach of regulatory authorities.

Finally, agile management methods are also utilised in MDDs. However, there are fewer case studies on this topic as it was primarily employed in software development. Nevertheless, medical software development projects managed via Agile have received approval under FDA and ISO regulations (Rasmussen et al. 2009; Duque and Kokol 2022). Additionally, Agile is gaining acceptance in projects involving product development ("Status Quo (Scaled) Agile 2020" n.d.). According to the law, the project management methodology must adhere to three principles to ensure product safety: effective risk management, quality management, and engineering. Agile project management addresses these three principles in a way that surpasses traditional waterfall project management. Lean and its hybrid versions also address these three principles. However, Agile demonstrates superior adaptability to the MedTech MDD process and has proven its effectiveness in developing complex products that integrate both software and hardware.

6.4 THE CASE STUDY IMPLEMENTED METHODOLOGY AND RESULTS

The development of the TRL 7 prototype involved orchestrating a diverse team consisting of eight specialists, each assuming roles aligned with the scrum team structure. Operating within the framework of Agile project management, this cohesive team, spanning a 16-month period, seamlessly incorporated HFE or Design for Usability (DfU) methodologies. The application of these methodologies unfolded seamlessly across both the conceptualisation and detailed design phases. Embracing a dynamic approach to continuous improvement, strategic checkpoints were strategically positioned at three sprints throughout the project, contributing to the iterative enhancement of the prototype. The proactive utilisation of DfU played a pivotal role in identifying and rectifying design flaws and usability concerns well before the validation phase commenced.

6.4.1 Implemented Management Methodology

To illustrate the management strategy, the following roadmap in Figure 6.3 details the Agile Development of Medical Devices (ADmed) main structure proposed by Martens et al. (2022). The roadmap in Figure 6.3 describes the five main phases of a MedTech project: initialisation, concept phase, detailed design, verification, validation, and release. The last two usually happen almost simultaneously. Therefore, they are represented in parallel. In the context of Figure 6.3, the conceptual phase is denoted as "project design," while the subsequent detailed design phase takes on the label of "realisation."

Figure 6.4 shows a detailed view of the ADmod model adapted to the case study presented in this work. The phases covered in this project are the initialisation, concept, detail design, and verification phases with their defined activities within each stage, as shown in Figure 6.4. It can be noticed that the project ends in the verification phase, as usually expected in an MDD.

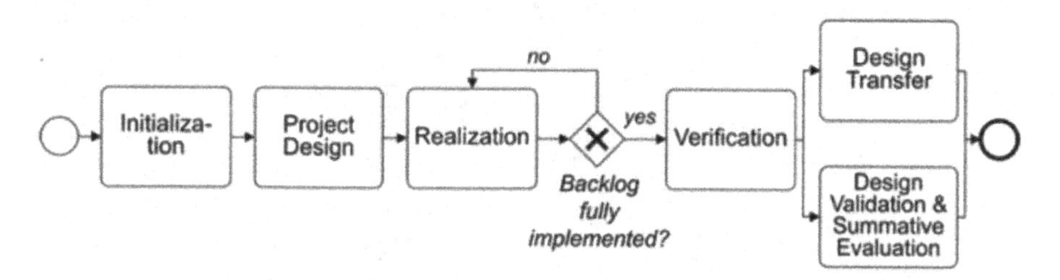

FIGURE 6.3 Main structure of the ADmed model (Martens et al. 2022).

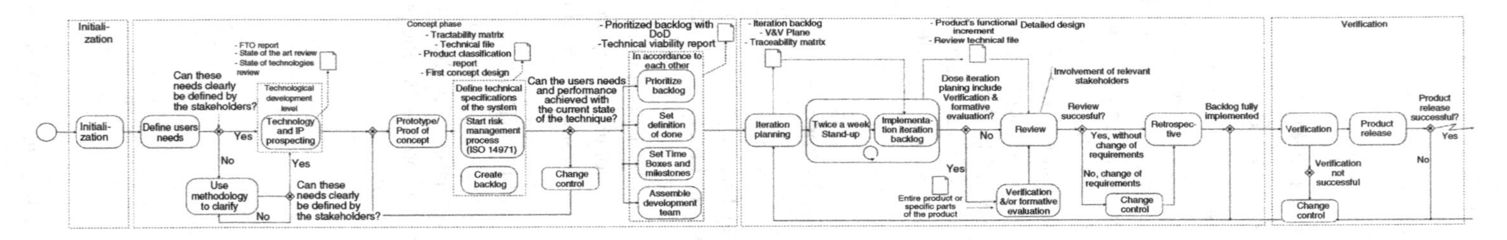

FIGURE 6.4 Extended view of the ADmed roadmap adapted by Casas et al. (2023).

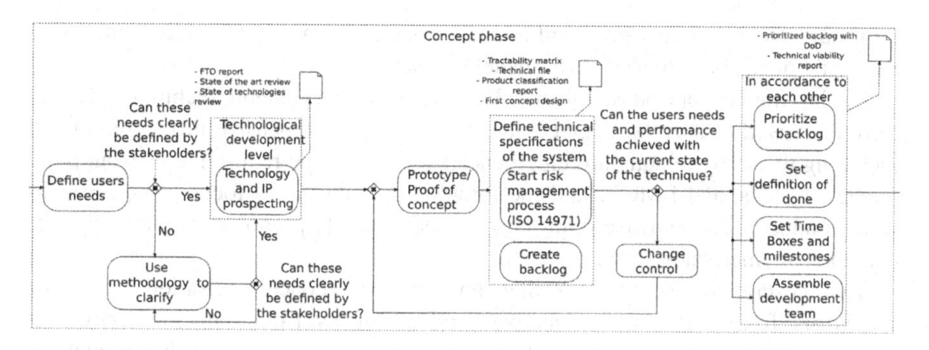

FIGURE 6.5 Concept phase by Casas et al. (2023).

6.4.2 Concept Phase (Sprint One)

As illustrated in Figure 6.5, the concept phase is characterised by two pivotal checkpoints. Upon embarking on the initial sprint, the primary objective was to establish a robust proof of concept prototype that aligned with the insights gathered during the requirements acquisition phase. This sprint marked a critical juncture in our Agile development process, where the focus extended beyond mere conceptualisation to the tangible creation of a prototype.

In Figure 6.6, one can observe that in the three-stage requirement-acquisition structure, we have first applied a learning stage to identify the problems/necessities of the emergency health service and their work environment and set three improvement hypotheses. The predefined hypothesis is to: (1) improve service by reducing care time, (2) reduce risks associated with ergonomics, and (3) achieve reliable and robust contactless measurements. Next, in the implementation stage, we worked with the emergency health service staff to develop a concept design and set the technical requirements to build a minimum viable product. Last, we validated that the

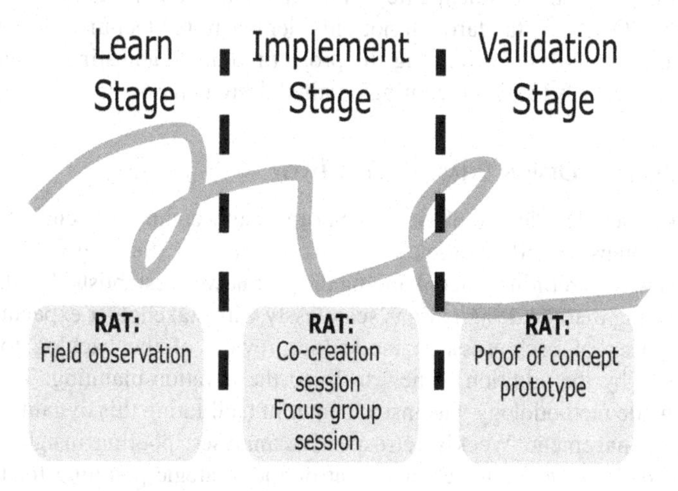

FIGURE 6.6 Three-stage requirement-acquisition structure.

minimum viable product provided enough evidence to answer the question, "¿Can the users' needs be achieved with further development?"

During the concept phase, diverse Requirements Acquisition Techniques (RAT) were employed to ascertain stakeholder needs. A comprehensive field observation, including the study of standard emergency procedures and requirements from paramedics, EMT's, and EMC's, laid the foundation. Subsequent focus group sessions were conducted to prioritise requirements and generate proposals aligned with the demands of emergency service agents.

Two co-creation sessions further enriched the conceptualisation, involving specialists from different perspectives and end-users from the emergency sector. The collaboration aimed at refining the design of the camera device and the interface of the associated application (APP). The iterative nature of Agile methodology allowed us to seamlessly integrate feedback, resulting in the generation of initial design drafts, material acquisition requirements, and Essential Design Outputs (EDOs).

In parallel, a Technology Acceptance Model (TAM) survey involving 120 participants provided valuable insights into the usability, functionality, and acceptance levels of the proposed technologies. This dual approach, combining qualitative insights from stakeholders and quantitative data from the TAM survey, ensured a comprehensive understanding of user needs and expectations (Marangunić and Granić 2015; Garmer et al. 2004). Although there are only two checkpoints implemented in the conceptual map as shown in Figure 6.5, the requirement-acquisition structure follows the three-stage approach mentioned in Section 6.3.

At this juncture, a Freedom to Operate (FTO) analysis, along with a thorough review of the state-of-the-art and state-of-technology, was conducted. These analyses provided a strategic backdrop for further development, ensuring alignment with existing technological landscapes and legal considerations.

The output of this sprint was not only a proof of concept prototype but also the groundwork for subsequent phases. The prototype laid the foundation for the Product Classification Report, Technical File Draft, and Traceability Matrix, all in accordance with ISO 13485 standards. It not only demonstrated technical feasibility, but also served as a catalyst for refining the project roadmap, adjusting schedules, and realigning team roles for the upcoming detailed design phase.

6.4.3 Detailed Design Phase (Sprint Two)

Entering the detailed design phase, the Scrum team, comprising eight specialists, engaged in intensive collaboration sessions twice a week. See Figure 6.7. The primary focus was on refining and enhancing the groundwork established in the concept phase, ensuring that the design aligns seamlessly with stakeholder expectations.

At the onset of each week, a meticulous review of the backlog took place, accompanied by the addition of new tasks to the iteration planning. The iterative nature of Agile methodology was instrumental in facilitating this dynamic approach to project management. Weekly retrospective analyses, pooling insights from the team, provided an avenue for change control and strategic planning for the subsequent iterations.

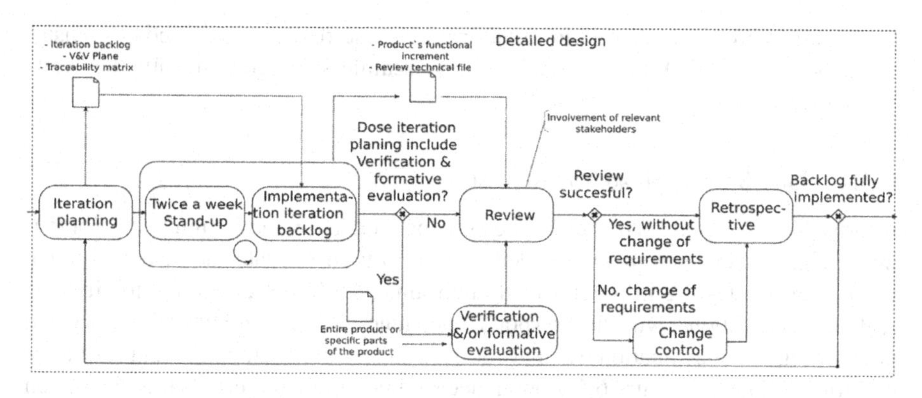

FIGURE 6.7 Detailed design phase by Casas et al. (2023).

During the second sprint, the team directed its efforts towards the development of a Programmable Logic Controller (PLC) and casing design, with the iteration loop represented in Figure 6.7 as a box that contains the twice-a-week focus group work session. The goal was to create a dummy prototype that closely mimicked the envisaged size and shape outlined in the initial design drafts. This dummy prototype, the hardware part, is depicted in Figure 6.8, which served as a crucial element in formative evaluations, allowing for a preliminary assessment of factors such as size, shape, and anticipated weight.

The evaluation encompassed aspects crucial to real-world deployment, including proposals for different attachment options on the body or backpack of paramedics and transport nurses during emergency services. Additionally, a heuristic analysis was conducted to scrutinise the handballing and grip of the device, ensuring optimal ergonomics and user acceptance.

The outcomes of this sprint played a pivotal role in informing subsequent development phases. The Product Classification Report, Technical File, and Traceability Matrix were revised and updated in tandem with the evolving design. The scrum

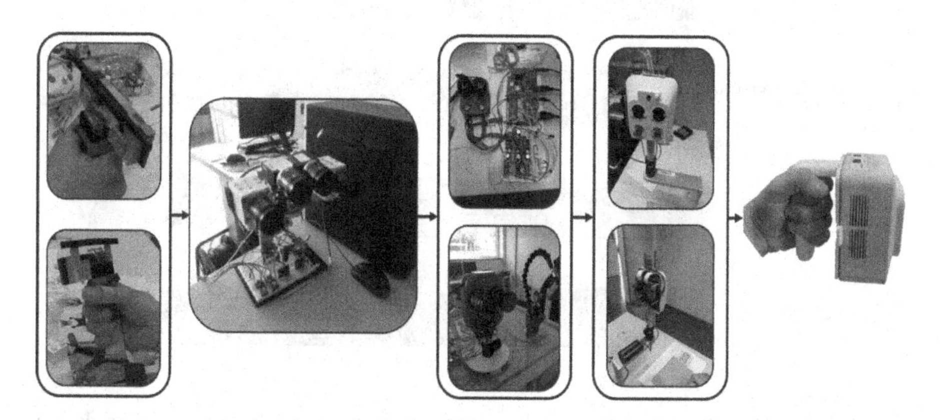

FIGURE 6.8 Development path of the hardware.

master assumed a central role in orchestrating these revisions and updates, ensuring adherence to ISO 13485 standards and seamless integration with the Agile framework.

6.4.4 Last Sprint before the Release

Approaching the final sprint before the anticipated release, the Scrum team, consisting of eight specialised members, dedicated efforts to validate the system through comprehensive testing in a controlled environment. This phase aimed to simulate real-world conditions, with first responders conducting trials in both laboratory and field settings. The evaluation was part of the system's pre-verification and formative usability evaluation. Figure 6.7 shows a decision point that represents this evaluation in the diagram.

During this sprint, the prototype, reaching a Technology Readiness Level (TRL) of 7, underwent rigorous testing to ensure its functionality and reliability. Various requirement acquisition techniques were employed to collect feedback from first responders, utilising a diverse set of data acquisition methods. This included insights gathered through simulations, ensuring the prototype's responsiveness to different scenarios.

The Agile project management approach facilitated continuous improvement throughout the project, with strategic checkpoints contributing to the refinement of the prototype. The multifunctional team collaborated seamlessly for 16 months, emphasising iterative development and incorporating user feedback at every stage.

Results from the last sprint, as shown in Figure 6.9, underscored the prototype's robustness, as it underwent thorough scrutiny in laboratory and field trials. The valuable data acquired during this phase played a pivotal role in refining design elements and addressing any remaining usability issues. The results of this evaluation show an

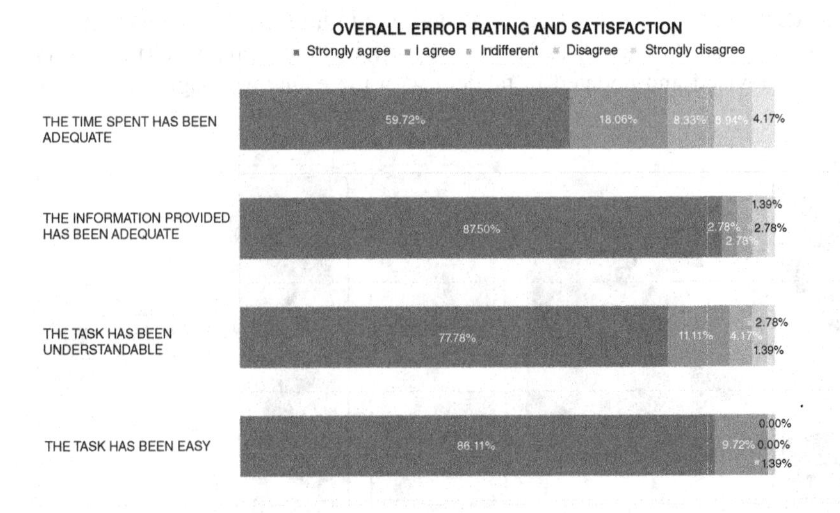

FIGURE 6.9 Overall error rating and satisfaction on the usability of the camera device and APP.

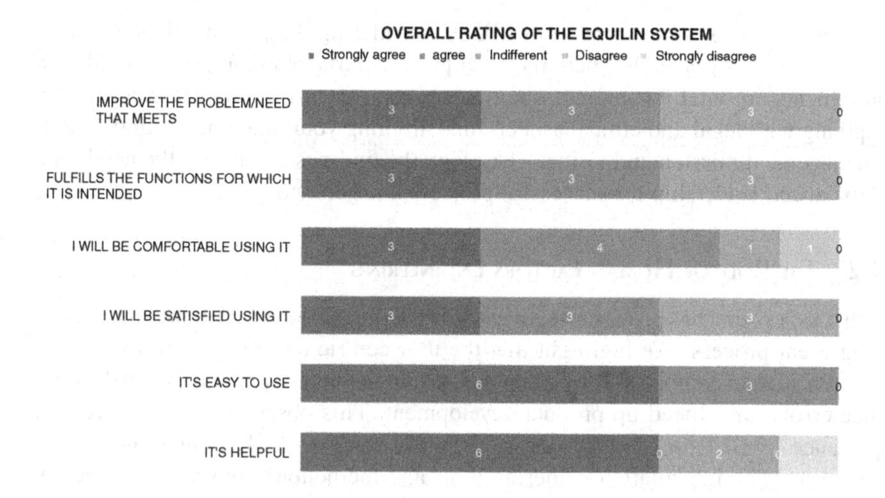

FIGURE 6.10 Overall rating and acceptance of the non-contact vital signs measurement system.

overall 90% success rate in the execution of the tasks to use the device, the information provided, and the information provided to make the equipment work. Here, the time spent was only 86% positively scored, as the APP had minor connectivity issues during the pairing process. However, the connectivity issues could be fixed in the detailed design phase.

On the other hand, the overall rating of the technological acceptance of the non-contact vital signs measurement system developed in the Equilin project also scored with a 90% acceptance rate, as seen in Figure 6.10.

This iterative approach not only prevented potential design flaws, but it also contributed to the reduction of development time by aligning with ISO 13485 standards without compromising design traceability and quality requirements.

6.5 DISCUSSION

This chapter outlines the challenges and principal management methodologies in the MedTech industry, with enfaces in medical device development (MDD). This section aims to develop more deeply into the implication of the findings and their significance in MedTech project management. Here, four implication aspects will be lined out: balancing regulatory compliance and user-centric innovation. Second, the role of human factors engineering (HFE). Third, the implication of implementing an iterative approach in project management. Last, regulatory compliance and standardisation.

6.5.1 Balancing Regulatory Compliance and User-Centric Innovation

Balancing regulatory compliance and user-centric innovation was one of the main targets of the leading figures of the project. Rigorously balancing those aspects is crucial, as an overemphasis on regulatory requirements can lead to project delays and increased costs in the context of MDD, considering that, as mentioned earlier, the product's life cycle ends where the MDI project starts. However, some regulation requirements are vital for maintaining traceability, offering good customer service, complying with legal and ethical aspects, and aligning your outcomes with the quality management mission and vision. Therefore, the findings emphasise the need for a well-balanced leadership for successful project management.

6.5.2 The Role of Human Factors Engineering

Another aspect that has proven effectiveness is the integration of HFE into the project management process. We highlight that the user-centric design approach shown in these case studies combined with Agile methodologies can improve user satisfaction, reduce errors, and speed up product development. This observation underlines the importance of HFE in MedTech projects in the context of MDD and encourages its adoption. The information-gathering tools and methodologies were strategically placed in the concept phase of the case study. As soon as the first tangible version of the hardware or the software was available, it was tested by users and/or experts. Even though the tested versions were not wholly functional at that point, it has been observed that they are still helpful tools to gather valuable information before moving further in development.

6.5.3 Implication of Implementing an Iterative Approach in Project Management

In the chapter, several other project management methodologies, each with its own characteristics, have been introduced and backed up with relevant state-of-the-art reviews. Although traditional methods like Stage-Gate or Concurrent Engineering have their place, iterative approaches in project management, such as Lean, Lean Six Sigma, and Agile, offer more flexibility in the context of MDD. The ability to iterate, adapt throughout the project life cycle, and combine HFE has demonstrated its advantages in addressing users' needs and ensuring product quality in the case study.

6.5.4 Regulatory Compliance and Standardisation

Last, the case study also shows that regulatory compliance and standardisation inside ISO 13485 is essential and possible. It is worth noting that regulatory agencies and organisations like the FDA recognise the value of Agile and Lean methodologies when applied property (Duque and Kokol 2022; McGrane et al. 2022; Huusko et al. 2023), especially in medical software development if the MedTech community does not start to embrace the importance of continuous improvement management methodologies combined with HFE. This recognition could lead to a smoother regulatory transition process and standardisation of good practices in the MedTech project management community. Besides, it provides evidence of a strategic implementation of user-centred design throughout more case study-oriented research.

6.6 IMPLICATION AND FURTHER DIRECTIONS

The implication of this research has enlightened throughout a case study that by embracing user-centric, Agile, and Lean methodologies, companies can reduce time-to-market, minimise errors, and ensure products align closely with user needs. Adopting HFE and iterative approaches will likely become more prevalent as companies strive to enhance their product development processes. In addition, the findings offer insights into potential improvements in regulatory compliance, which could lead to a more streamlined and efficient validation process for MedTech products.

6.7 CONCLUSION

In conclusion, this case study presents a novel approach to HFE and Agile methodology in the development of a medical device. The project, spanning 16 months and adopting an Agile product development model, successfully navigated through three sprints, culminating in the creation of a non-contact vital signs measuring system.

A noteworthy aspect of this study is the integration of HFE and Agile to overcome the limitations associated with traditional Stage-Gate project management in the medical device domain. Throughout the concept and detailed design phases, DfU tools played a crucial role in generating value. These tools, implemented in parallel with strategic considerations such as IP strategy and risk management, ensured a user-centric approach.

The collaboration of a multifunctional team organised according to the Scrum model proved instrumental in the iterative development process. The last sprint involved rigorous testing of the prototype, involving first responders and achieving a TRL of 7. Results from laboratory and field trials were continuously integrated into the project, contributing to the refinement of the prototype and scored with over 90% error free evaluation and technological acceptance by the professionals who took part in the formative evaluation during the last sprint.

Importantly, this Agile and HFE approach not only expedited the development timeline, but also upheld the necessary design traceability and quality standards outlined in ISO 13485. The flexibility of Agile, combined with the focus on Human Factors, demonstrated the potential to minimise errors, enhance user satisfaction, and streamline the development of complex medical devices.

ACKNOWLEDGEMENTS

The authors extend their gratitude to the European Regional Development Fund (ERDF) through the Pluri-regional Operational Program of Spain (POPE) 2014–2020 and the Ministry of Science and Innovation for their financial support in the form of the FID programme, "Promotion of Innovation from Demand." Special appreciation goes to the coordination efforts of the 061 Health Emergencies Centre of the Andalusian Health Service in providing invaluable access to their facilities and insights from emergency service professionals.

The collaborative spirit of our colleagues from Plux, Pedro Duque, and Goçalo Telo significantly contributed to the success of the EQUILIN project. We also acknowledge the insightful contributions of our colleagues from IBV: Úrsula Martinez, Juanma Belda, Daniel Gómez, Cristina Díaz, David Rubio, David Moro, Amparo Lopez, and Carlos Chirivella. Their expertise played a crucial role in advancing this research.

This acknowledgment recognises the collective efforts of all individuals and entities involved, emphasising their indispensable role in shaping the outcomes of this research endeavour.

REFERENCES

Anderson, Kelly M., Scott E. Grasman, Kamal Ayoub, Steve Introne, and Kevin Smithwick. n.d. Using Lean Product Development to Speed Time to Market for Medical Devices. https://www.proquest.com/scholarly-journals/using-lean-product-development-speed-time-market/docview/1190411357/se-2?accountid=28445. Accessed 6 November 2023.

Boylan, Brian, Olivia McDermott, and Niall T. Kinahan. 2021. Manufacturing Control System Development for an In Vitro Diagnostic Product Platform. *Processes* 9(6): 975. https://doi.org/10.3390/pr9060975.

Byrne, Brian, Olivia McDermott, and John Noonan. 2021. Applying Lean Six Sigma Methodology to a Pharmaceutical Manufacturing Facility: A Case Study. *Processes* 99(3): 550. https://doi.org/10.3390/PR9030550.

Casas, Adrian M., Amparo L. Vicente, Lorenzo Solano-García, and Jose Laparra. 2023. 3 Sprints from Zero to Innovative Medical Device in 16 Months: Benefits of Combining Human Factors and Agile. *Production Management and Process Control* 104(2023): 8–18. https://doi.org/10.54941/AHFE1003047.

Cooper, Robert G., and Anita F. Sommer. 2016. The Agile-Stage-Gate Hybrid Model: A Promising New Approach and a New Research Opportunity. *Journal of Product Innovation Management* 33(5): 513–526. https://doi.org/10.1111/JPIM.12314.

De Rossi, Danilo. 2012. An Overview of Lean and Six Sigma a Framework for Organisational Development ! https://www.academia.edu/11660265/An_Overview_of_Lean_and_Six_Sigma_A_Framework_for_Organisational_Development.

Duque, Redondo, and Peter Kokol. 2022. Agile Software Development in Healthcare: A Synthetic Scoping Review. *Applied Sciences* 12(19): 9462. https://doi.org/10.3390/APP12199462.

Fairbanks, Rollin J., and Stanley Caplan. 2004. Poor Interface Design and Lack of Usability Testing Facilitate Medical Error. *The Joint Commission Journal on Quality and Safety* 30 (10): 579–584. https://doi.org/10.1016/S1549-3741(04)30068-7.

FDA. n.d. Design Control Guidance for Medical Device Manufacturers. https://www.fda.gov/regulatory-information/search-fda-guidance-documents/design-control-guidance-medical-device-manufacturers. Accessed 9 March 2022.

Garmer, Karin, Jessica Ylvén, and I. C. Mari Anne Karlsson. 2004. User Participation in Requirements Elicitation Comparing Focus Group Interviews and Usability Tests for Eliciting Usability Requirements for Medical Equipment: A Case Study. *International Journal of Industrial Ergonomics* 33(2): 85–98. https://doi.org/10.1016/J.ERGON.2003.07.005.

Gerber, Carlos, Kristin Goevert, Sebastian Schweigert-Recksiek, and Udo Lindemann. 2019. Agile Development of Physical Products: A Case Study of Medical Device Product Development. In Chakrabarti, A. (ed.) *Research into Design for a Connected World. Smart Innovation, Systems and Technologies*, vol. 135, pp. 823–34. Springer Science and Business Media Deutschland GmbH, Singapore. https://doi.org/10.1007/978-981-13-5977-4_69.

Goldenberg, Seth J., and Jeff Gravagna. 2018. A Real-World Perspective: Building and Executing an Integrated Customer Engagement Roadmap That Bridges the Gaps in Traditional Medical Device Development Processes. *Journal of Medical Marketing*, 16 (2): 41–19. https://doi.org/10.1177/1745790418770598.

Huusko, Juhamatti, Ulla M. Kinnunen, and Kaija Saranto. 2023. Medical Device Regulation (MDR) in Health Technology Enterprises: Perspectives of Managers and Regulatory Professionals. *BMC Health Services Research* 23(1): 1–12. https://doi.org/10.1186/S12913-023-09316-8/TABLES/8.

Khan, Muhammad S., Ahmed Al-Ashaab, Essam Shehab, Badr Haque, Paul Ewers, Mikel Sorli, and Amaia Sopelana. 2013. Towards Lean Product and Process Development. *International Journal of Computer Integrated Manufacturing* 26(12): 1105–1116. https://doi.org/10.1080/0951192X.2011.608723.

Lin, Laura, Kim J. Vicente, and D. John Doyle. 2001. Patient Safety, Potential Adverse Drug Events, and Medical Device Design: A Human Factors Engineering Approach. *Journal of Biomedical Informatics* 34(4): 274–284. https://doi.org/10.1006/JBIN.2001.1028.

Lumsdaine, Edward, and Monika Lumsdaine. 1994. Creative Problem Solving. *IEEE Potentials* 13(5): 4–9. https://doi.org/10.1109/45.464655.

Marangunić, Nikola, and Andrina Granić. 2015. Technology Acceptance Model: A Literature Review from 1986 to 2013. *Universal Access in the Information Society* 14(1): 81–95. https://doi.org/10.1007/S10209-014-0348-1/TABLES/3.

Marešová, Petra, Blanka Klímová, Jan Honegr, Kamil Kuča, Wan N. H. Ibrahim, and Ali Selamat. 2020. Medical Device Development Process, and Associated Risks and Legislative Aspects-Systematic Review. *Frontiers in Public Health* 8(July): 308. https://doi.org/10.3389/FPUBH.2020.00308.

Martens, Maren, Anna Schidek, Markus Schmidtner, and Holger Timinger. 2022. ADmed: An Adaptive Technical Process for the Agile Development of Medical Devices. In *Proceedings of the 14th International Joint Conference on Knowledge Discovery, Knowledge Engineering and Knowledge Management*, vol. 3: IC3K, pp. 177–184. SCITEPRESS - Science and Technology Publications, Valletta, Malta. https://doi.org/10.5220/0011543100003335.

McGrane, Vincent, Olivia McDermott, Anna Trubetskaya, Angelo Rosa, and Michael Sony. 2022. The Effect of Medical Device Regulations on Deploying a Lean Six Sigma Project. *Processes* 10 (11): 2303. https://doi.org/10.3390/PR10112303.

MedTech Europe. n.d. MedTech Europe's Facts and Figures 2022. https://www.medtecheurope.org/resource-library/medtech-europes-facts-and-figures-2022/. Accessed 8 February 2023.

Mitchell, Rebecca J., Ann Williamson, and Brett Molesworth. 2015. Use of a Human Factors Classification Framework to Identify Causal Factors for Medication and Medical Device-Related Adverse Clinical Incidents. *Safety Science* 79(November): 163–174. https://doi.org/10.1016/J.SSCI.2015.06.002.

Neagoe, Lavinia N., and Vladimir M. Klein. July 2010. Quality and Management Tools, An Integrated Approach for Quality Cost Reduction. *Revista Recent* 11(2)(29), Romania.

Pietzsch, Jan B., Lauren A. Shluzas, M. Elisabeth Paté-Cornell, Paul G. Yock, and John H. Linehan. 2009. Stage-Gate Process for the Development of Medical Devices. *Journal of Medical Devices* 3(2): 021004 (15 pages). https://doi.org/10.1115/1.3148836.

Rasmussen, Rod, Tim Hughes, J. R. Jenks, and John Skach. 2009. Adopting Agile in an FDA Regulated Environment. *Proceedings-2009 Agile Conference, AGILE 2009,* Chicago, IL, pp. 151–55. https://doi.org/10.1109/AGILE.2009.50.

Rathi, Rajeev, Ammar Vakharia, and Mohd Shadab. 2022. Lean Six Sigma in the Healthcare Sector: A Systematic Literature Review. *Materials Today: Proceedings* 50: 773. https://doi.org/10.1016/J.MATPR.2021.05.534.

Roma, Marylene S. G., and Euler de Vilhena Garcia. 2020. Medical Device Usability: Literature Review, Current Status, and Challenges. *Research on Biomedical Engineering* 36: 163–170. https://doi.org/10.1007/s42600-019-00037-8.

Salleh, Narishah M., and Puteri N. E. Nohuddin. 2019. Comparative Study between Lean Six Sigma and Lean-Agile for Quality Software Requirement. *International Journal of Advanced Computer Science and Applications* 10(12): 212–218. https://doi.org/10.14569/ijacsa.2019.0101230.

Saidi, T., C. T. Mutswangwa, and T. S. Douglas. 2019. Design Thinking as a Complement to Human Factors Engineering for Enhancing Medical Device Usability. *Engineering Studies*, 11(1): 34–50. https://doi.org/10.1080/19378629.2019.1567521.

Slattery, Owen, Anna Trubetskaya, Sean Moore, and Olivia McDermott. 2022. A Review of Lean Methodology Application and Its Integration in Medical Device New Product Introduction Processes . *Processes* 10(10): 2005. https://doi.org/10.3390/pr10102005.

Sloan, Kenneth J. 1996. Calhoun: The NPS Institutional Archive Leveraging Management Improvement Techniques. *Management Review, Reprint Series* 38(1): 69–79. https://hdl.handle.net/10945/43834.

Status Quo (Scaled) Agile 2020. n.d. https://www.hs-koblenz.de/bpm-labor/status-quo-scaled-agile-2020. Accessed 6 February 2023.

Sun, X., R. Houssin, J. Renaud, and M. Gardoni. 2019. A Review of Methodologies for Integrating Human Factors and Ergonomics in Engineering Design. *International Journal of Production Research* 57(15–16): 4961–4976. Taylor & Francis Ltd. https://doi.org/10.1080/00207543.2018.1492161.

Van der Peijl, J., J. Klein, C. Grass, and A. Freudenthal. 2012. Design for Risk Control: The Role of Usability Engineering in the Management of Use-Related Risks. *Journal of Biomedical Informatics*, 45(4): 795–812. https://doi.org/10.1016/J.JBI.2012.03.006.

World Health Organization. 2010. *Medical Devices : Managing the Mismatch : An Outcome of the Priority Medical Devices Project.* World Health Organization, Switzerland, p. 147. ISBN: 978 92 4 156404 5.

7 User Context of Enterprise Information Systems for Manufacturing

Krzysztof Hankiewicz

7.1 INTRODUCTION

The development of information systems for enterprises has always depended on the development capabilities of computer hardware, software and network infrastructure, as well as on the staff who would be able to use these systems. Initially, the human factor was not appreciated as an essential element of the proper functioning of the IT system. Human limitations were also not taken into account at the software design stage, assuming that appropriate training would be sufficient. Although those times are gone, the complexity of current systems requires further improvement of human–computer dialogue.

IT systems used in production processes are among the most complex systems in enterprises. In many cases in the past, when a functioning company implemented an IT system, the production module was implemented last, even with a delay of several years in relation to all other modules of the company's information system. Nowadays, it is difficult to imagine production systems without full IT support. The ability to quickly respond to changing customer needs often depends on its efficiency.

In the times of Industry 4.0 and AI, it is difficult to imagine a company's production department that is not related to the enterprise's information system. However, the complexity of these interconnected systems continues to increase. This increases the possibilities of their use in the enterprise. However, there is a concern that the excessive complexity of such a system may make them difficult to use. As a result, this may affect the efficiency of using the designed software, for example, by affecting the response time and operator errors.

When we go back to the early stages of IT development, we can say that already at the stage of supporting office work, attention was paid to the user's context, noting the benefits of adapting the software to human mental capabilities, in the form of reducing the number of errors made. At this stage, it was difficult to talk about the intuitiveness of the software. However, for example, efforts were made to standardize the menu layout. In the case of more complex systems, including enterprise information systems,

DOI: 10.1201/9781003505327-7

these activities were usually insufficient. This made it necessary to conduct long employee training sessions. The use of a graphical interface allowed us to develop the concept of intuitive software and even introduce visualization of certain processes. Moreover, it allowed for better control of business and production processes.

Collecting more data opens up new possibilities, but also requires processing a huge amount of data in the final phase of analysis and human decision-making. With the use of AI, a human can also be supported at the stage of data analysis and decision-making, but there is always the need to supervise processes. Therefore, we are still dealing with human–system interaction, and the responsibility for the decisions made usually does not decrease, although in some cases you can count on their verification based on the knowledge base.

Many software development companies take usability requirements into account at the design stage, focusing mainly on the graphical user interface. Often, it is also necessary to adapt the layout of menus, icons and other interface components to other programs. These do not always provide sufficient ease of use. The use of production systems still usually requires long employee training. However, training does not completely prevent errors and excessive workload when the interface is complex and non-intuitive.

The aim of the research is to analyse the conditions for using the information system for manufacturing in a selected company. The analysis focused on features specific to systems supporting production processes. Selected examples of systems were analysed based on software inspections in the field of ergonomics and interviews with employees. In order to obtain greater freedom of probing with respondents, semi-structured interviews were used. Respondents related to the operation of the production system were selected. Where possible, interviews continued, prompting respondents to talk normally and comment on difficulties in using the production system. The aim of the research was not to maximize the research sample, but to collect as many categories of problems as possible encountered by operators using the analysed system. For this reason, the operators not only answered the prepared questions, but also had the opportunity to speak freely during the interviews.

7.2 INFORMATION SYSTEMS FOR MANUFACTURING

Information systems designed to support production are intended to comprehensively support key areas related to the organization of the production process. It is primarily planning, management and accounting of production. Their use should significantly increase work efficiency, preview and analysis of the current production situation and production orders. In addition, they are intended to help in meeting production order deadlines.

The assumption is that Information Systems for Manufacturing should be characterized by a friendly interface, integration and data exchange with ERP systems, and the result of their work should be a shorter time of production operations. For example, an MES (Manufacturing Execution System) class system is a production management system that allows to collect data from machines or the entire production line, analyse and display it in a user-friendly way. MES software is being implemented to make operators' work more predictable and help managers make more insightful decisions for faster and better production. Such a system offers information and a comprehensive overview of production parameters, which is useful in identifying inefficient processes and cause-and-effect shortcomings. This in turn enables the correction of underlying problems, leading to increased efficiency and better allocation of resources. Moreover, these systems aid in quality control, making it easier to maintain compliance with industry standards and regulations. By tracking and documenting the production process, these systems minimize product defects, taking into account all production units. They also improve communication within the organization and help reduce downtime, making production much more efficient. It is also important to remember the complexity of these systems in large manufacturing enterprises. Bratukhin and Sauter argue that centralized MES applications cannot adequately cope with unpredictable order flow and changes on the shop floor. Therefore, they propose transferring part of the system's functionality to the field level, which would provide a more flexible production system (Bratukhin and Sauter, 2011). In turn, for small- and medium-sized companies, the financing of the system itself may be a problem. Therefore, it is often proposed to implement an integrated system containing a production module (Pfeifer, 2021).

The expectations towards Information Systems for Manufacturing are primarily:

- real-time data flow and support for decision-making processes;
- easier identification of real bottlenecks in the production process;
- verification of the company's possession of the resources necessary for production;
- reading production data directly from machines in the form of production counters, machine status, process parameters, allowing you to control the efficiency of machines;
- data caching in the event of loss of connection to machines to retain all information;
- analysis of data so that it can be presented to the operator or manager in a way that allows immediate decisions to be made;
- better traceability of production processes and faster detection of production defects;
- documentation flow in digital form; consequently, all documents are stored in the system, which allows version control and access levels to be defined;
- constant quality monitoring: process parameters are collected automatically from the automation level or manually from operators, including measurements and checklists;
- production visualization and display of production progress and current delay for ongoing production;
- immediate reporting of failures along with their cause and comments, as well as information about the failure in the maintenance department; and
- integration with external systems.

In addition, other aspects are as follows:

- monitoring and controlling the quality of the production process using statistical methods – SPC (Statistical Process Control);
- monitoring key performance indicators for individual machines (Mean Time To Repair, Mean Time Between Failures and Mean Time To Failure);
- resource scanning to match components and tools for production;
- scanning of components and materials used in production, along with the ability to link which material was used in the production of a specific product or its batch;
- easier control of unit production, as well as processing previously produced elements (defective elements) by tracking and sending the defective element to the appropriate machine; and
- modelling of basic production data such as positions, routes and factory structure.

Manufacturing Execution System software can help identify many inappropriate processes occurring from the production floor to the warehouse, which can be eliminated or corrected, for example, by simplifying procedures or eliminating documentation forms that have been identified by production reports as a factor causing disruptions. Step by step, making small changes can add up to significant improvement potential and enable an increase in OEE (Overall Equipment Effectiveness). Very often, the Manufacturing Execution System allows manufacturing companies to save money by saving time in various production processes. It is also possible, in some cases, to reduce the burden on employees in production as well as the surrounding work areas involved in inter-site processes. Additionally, incorporating elements of artificial intelligence can enable automation using machine learning models or logic-based systems (Zdravković et al., 2022).

7.3 ERGONOMIC FEATURES OF SOFTWARE THAT AFFECT THE USER CONTEXT

The growing amount of information is a problem not only with its perception but also with access to information necessary (and important) in a given situation to make decisions. It is even claimed that the accumulation of information makes access to it difficult. This has been called the data availability paradox (Woods et al., 2002). Of course, the imbalance between the amount of incoming information and the human ability to process it is a source of stress. Difficulties related to the reception, processing and use of information (having more information than one can acquire, process, store or retrieve) are experienced as mental discomfort. Eppler and Mengis (2004) describe information overload in the following way: "Information overload occurs when the supply exceeds the capacity. Dysfunctional consequences... and a diminished decision quality are the result". Researchers from various disciplines have found that a person's performance correlates positively with the amount of information they receive up to a certain point. Beyond this point, individual's performance declines rapidly, as described by an inverted U-shaped curve (Chewning and Harrell, 1990).

Information overload is believed to be so important because information is arguably the most valuable human asset in a knowledge economy (Hemp, 2000). A person who is in a situation of information overload is aware of this, but does not always have the influence to change this state. Therefore, it is necessary to supervise from the outside whether the amount of incoming information is appropriate to the perceptual capabilities of employees.

When analysing the user context during human–computer interaction (HCI), human limitations must be taken into account. These include (Sikorski, 2010):

- visual limitations (adjusting the presentation of objects on the screen so that they can be read correctly),
- limitations related to memory capacity and the ability of the operator to process information (adapting the method of encoding information),
- limitations on manipulation skills (by adjusting the size of manipulation elements and access to them so that they do not excessively burden the operator's muscles),
- ensuring comfort (adapting working conditions such as lighting, microclimate, noise and working position to the operator),
- ensuring the appropriate pace of work – preferably with the possibility of adapting it,
- counteracting fatigue – among others, by introducing breaks at work,
- counteracting stress caused by working conditions and the excess of tasks to be performed.

Based on the analysis of the user's context and technical capabilities, functions and tasks should be divided between the user and the system. Therefore, the basis for developing a description of the needs and requirements for the system will be the users' predispositions.

The area of issues discussed in the article is related to the ergonomic features of information systems. These features determine, among other things, whether the software takes into account human perceptive capabilities and whether it is a source of excessive stress at work. In this case, it is essential to analyse the functional characteristics of the entire system that affect the work of operators.

If we analyse the ergonomic criteria, we can conclude that the decisive factor is easy and understandable operation of the software, and in its form accepted by employees (Perret et al., 2002). Software ergonomics focuses on adapting the designed software to humans. The ergonomically designed software helps:

- reduce stress,
- increase processing efficiency and
- improve IT acceptance.

For this reason, it is necessary to define the criteria (requirements) that should be taken into account when selecting software. The ISO 9241-110 standard identifies interaction principles important for the design and assessment of systems:

- Suitability for the user's tasks
- Self-descriptiveness
- Conformity with user expectations
- Learnability
- Controllability
- Use error robustness
- User engagement

Dialogues should be designed to support task completion, although the user is additionally burdened with the properties of the dialogue system (Krug, 2006). As an example, you can switch from the general view to the detailed view at any time without much effort (e.g., without changing the menu). In the case of a production system, it is possible to change the view of tracking another technological process or another phase of it when production is carried out in a pipeline. In other words, for a system to be suitable for the user's tasks, it must be based on the characteristics of the task performed by the user.

In terms of self-descriptiveness, the dialogue should be designed in such a way that each individual step of the dialogue is immediately understandable or the user can receive explanations of a given step of the dialogue upon request. For example, in the case of a widely branched menu, it should be possible to obtain information about its current status at any time.

Talking about conformity with user expectations, we can give an example of a dialogue box, which should be designed in such a way that it meets the expectations of users that they bring with them from the experiences that have developed during the previous use of dialogue systems. The condition is that the dialogue box and all help functions (including the user manual) are available in a language that the user understands. For example, if a user works with several software packages, the same command must be designed in the same way for different programs (uniform user interface).

When analysing the principle of learnability, it is usually emphasized that it minimizes the need for learning and provides support when learning is needed. For a production system, minimizing the need for learning seems to be quite a risky proposition due to the complexity of the system and the consequences of incorrect use. The solution seems to be the introduction of a system operation simulator option, which will allow not only to train new employees but also to test new possibilities when changing the production profile.

When it comes to controllability, the dialogue box should be designed in such a way that the user can influence the speed of the process and the selection and order of commands. For example, when correcting input data, it must be possible to correct individual values without having to go through all the values or even re-enter them.

Considering use error robustness, it can be concluded that the dialogue box should be designed in such a way that erroneous entries can be corrected with little effort. User input must not lead to undefined system states or system failures. For example, there should be a command that undoes the last step or more steps, e.g., when the wrong system function has been accidentally selected. In a production system, the production process should be launched after additional confirmation of the introduced settings. Additionally, such systems are equipped with tools that verify the compliance of selected functions.

The principle of user engagement can be considered relatively new, at least when it comes to introducing it into the norm. It implies that functions and information should be presented in an encouraging and motivating way, supporting continuous interaction with the system. In the case of a production system, a supporting element may be the visualization of the system's operation based on sensors monitoring the ongoing processes.

Very often, ergonomic software evaluation focuses on usability assessment, which is based on checking how users can use the product to achieve specified goals. Thus, it can be concluded that the usability evaluation is somewhat subjective, as it involves evaluating the interaction between the user and the product. To make the usability assessment more objective, evaluation criteria are defined and usability tests are performed. Jacob Nielsen points out the multidimensional nature of usability. For a product or service to be usable, it must include at least these five basic quality components (Nielsen, 1993, 2012):

- Learnability
- Efficiency
- Memorial
- Error tolerance and prevention
- Satisfaction

Usability can therefore be defined as "a quality attribute that assesses how easy user interfaces are to use" (Nielsen, 2012). The classic ISO 9241-11 standard defined usability as "extent to which a system, product or service can be used by specified users to achieve specified goals with effectiveness, efficiency and satisfaction in a specified context of use." Generally, this determines the level of effectiveness and satisfaction with which users can achieve their goals in a specific way (Bevan, 2001).

Usability is typically measured through a representative group of test users. Additionally, a specific set of tasks is prepared for test users or research is carried out using real users performing typical tasks. In both cases, the important point is that usability is measured with respect to specific users and specific tasks. It may well be the case that the same system will be assessed as having different performance properties if it is used by different users for different tasks. Usability measurement therefore begins with defining a representative set of test tasks against which various usability attributes can be measured.

When designing software, the expected reactions of users are increasingly taken into account. Such design is treated as an extension of User Interface design and is often even referred to as User Experience design. The important thing is that usability is not the same as user experience. According to Garett, User Experience is "the experience the product creates for the people who use it in the real world" (Garrett, 2011). Norman and Nielsen (2022) explain that: "User experience encompasses all aspects of the end-user's interaction with the company, its services, and its products." This term is not new, but has become more popular since 2000 (https://books.google.com/ngrams/). User Experience is defined by the ISO 9241-210 standard as:

"person's perceptions and responses resulting from the use and/or anticipated use of a product, system or service." It can therefore be said that User Experience has a wider scope of meaning than usability (Norman and Nielsen, 2022). This term describes what happens when using a given product, and how it affects the user's feelings and emotions.

In modern applications of computers in enterprises, there are few cases of individual work with the system. This may be the use of a computer for office work or individual design work. However, even in such situations, cooperation occurs in the performance of a specific task. It can be said that most of the joint work of teams is based on the enterprise's information system (Hankiewicz, 2012). This system provides not only the software necessary for work, but also communication between team members and methods of tracking progress of their work, as well as technological processes and other tasks being performed. Therefore, we are dealing here not only with the mutual influence of a person and a system, but also with the influence of a person through the system on another person. Changes in the system often require automatic notification to other employees. In the case of production systems, notifications should concern not only critical states, but also significant changes in production process parameters. In many cases, it is also expected that basic production parameters are available in the system and that they are archived so that their subsequent analysis is possible. When developing a production management system, attention should therefore be focused on users, rather than just technological processes. The principles of interaction design should be used, which was postulated many years ago (Preece et al., 1994, 2002).

7.4 RESEARCH

7.4.1 Research Methodology

The research was planned as a case study in a manufacturing company in the mechanical industry. They were carried out in two stages. In the first stage, user surveys were used, and in the second stage, interviews with employees were performed.

When selecting the survey evaluation method, the author's experience from previous research was used. Their effectiveness, complexity and reliability of results were taken into account.

Usability is traditionally associated with five usability attributes (compared with the information given in Chapter 3), but the list of these attributes may be expanded and more detailed. The author conducted research using various lists of attributes. One of them in previous studies (Hankiewicz and Prussak, 2005; Prussak and Hankiewicz, 2007) was the following:

Easiness of use
Usefulness
Comprehension
Using fastness
Accessibility
Self-descriptiveness

Adequateness
Error tolerance
Aesthetics
Ease learning of use
Integrity

The above attributes were treated as group criteria, while individual criteria that served as the direct basis for formulating detailed questions addressed to users were distinguished (there were over 40 of them in the conducted research).

The disadvantage of this research methodology was some inconvenience for respondents. A large number of detailed questions required great concentration. Moreover, when users additionally assessed the importance of elementary criteria, it turned out that for some, defining the degree of importance of a specific criterion requires more thought, while others define it on the spur of the moment.

Since one of the disadvantages of previously conducted research was the burdensomeness of surveys consisting of several dozen very precise questions, in this case, the System Usability Scale (SUS) method, significant in terms of popularity and period of use, was chosen, as an old industry standard proposed by John Brooke over 35 years ago (Brooke, 2013; Sauro, 2011). Instead of answering "yes" or "no," we can use five answer options: from "strongly agree" to "strongly disagree." Selected method turned out to be so universal, and even though it was developed at a time when IT systems had a completely different interface, it is still used. An example of the use of this method by the author was the usability study of an open ERP system (Hankiewicz and Jayathilaka, 2018).

A relatively simple method of initial assessment could have been chosen due to the continuation of research in the form of interviews with employees, during which they were to respond in detail to questions regarding selected aspects of the use of Information Systems for Manufacturing in their company.

In production systems, software usability assessment focuses on its functionality. User expectations are usually in the background. All the more so because the users themselves focus their attention on completing the task and, in many situations, do not analyse their interaction with the system. However, when interviewing employees, you can count on a retrospective assessment of these interactions. Employees remember what problems occurred during work, what types of mistakes they made and whether the mistakes were repeated in a similar sequence of using the system. Of course, you cannot expect clear answers. This is due to many factors, such as different experiences related to the previous use of other systems, the level of knowledge about the structure and interface of the system, the level of concentration on work or the level of fatigue. On the other hand, working at a higher level of fatigue may allow you to detect deficiencies in the intuitiveness of the interface. There are mistakes that employees usually do not make when they are focused and act according to the learned algorithm. It also happens that mistakes made when the level of fatigue is high are not remembered. Then, in addition to interviews, observing employees may be helpful. In each case, however, the purpose of analysing user expectations is to indicate the essential features of the interface or the entire system and directions for their improvement.

For the above reasons, semi-structured interviews were selected. The flexible structure of the interview allows the researcher to ask an additional question depending on the interviewee's statement when he or she needs more information in the context of the answer obtained. Semi-structured interviews also give informants the freedom to express their views in their own way. In general, it can be said that the semi-structured interview is a qualitative research method that combines a predetermined set of open questions (questions triggering discussion) with the opportunity for the interviewer to delve deeper into individual topics or answers. It can therefore be concluded that a semi-structured interview method combines some structured questions with some unstructured exploration (Wilson, 2010, 2014).

Usability testing with users, especially for such complex systems as production systems, may not reflect the full picture of its imperfections. Interviews with users allow you to spot problems that are difficult for users to categorize when completing a survey. Therefore, it can be concluded that user interviews provide greater opportunities for further software improvement.

7.4.2 Results from Surveys

As described earlier, survey research was the first stage of the research. They were conducted based on SUS Questionnaires. This was the stage preceding interviews with employees. People who use the production system in the studied enterprise were selected for the study. With this limitation defined, it was possible to identify 23 people who met this condition. These people were asked to complete the survey. The degree of use of this system by the respondents was not analysed, assuming that even short-term use allows one to form an opinion about its functioning. A summary of the obtained results is presented in Table 7.1.

The obtained results can generally be considered a positive assessment of the system. The simplicity of the study allows you to read the results directly, but it is worth

TABLE 7.1
Results Obtained from User Surveys

Questions	Mean	Median	Mode
I think that I would like to use this system frequently.	3.02	3	3.0
I found the system unnecessarily complex.	3.51	3	3.5
I thought the system was easy to use.	4.10	4	4.0
I think that I would need the support of a technical person to be able to use this system.	3.50	3	4.0
I found that various functions in this system were well integrated.	3.70	4	4.0
I thought there was too much inconsistency in this system.	2.42	2	3.0
I would imagine that most people would learn to use this system very quickly.	2.45	2	2.0
I found the system very cumbersome to use.	3.53	4	4.0
I felt very confident using the system.	2.10	2	2.0
I needed to learn a lot of things before I could get going with this system.	3.33	4	3.5

pointing out the elements that are related to each other. Respondents stated that the system is not too difficult to use, but implementation is relatively complicated and requires training in its operation. Moreover, the unnecessary complexity of the system used was pointed out, which seems particularly disturbing due to the potential consequences of errors made. Similarly, in the case of answers confirming that the system integration is appropriate, we also have answers regarding the inconsistencies that occur.

7.4.3 Results from Interviews

Interviews, and especially semi-structured interviews, are a method in which the results obtained are not suitable for statistical analyses, and it is even difficult to average what is obtained as a result of the interviews. For this reason, many researchers avoid this research method. However, in many cases, surveys do not show the broader context of the situation. Not to mention cases where the survey questions asked are incomprehensible or ambiguous to the respondent. The flexible structure of the interview allows for two-way communication and asking additional questions if necessary. Although based on interviews we do not obtain such an ordered list of answers as in the case of surveys, in the case of complex processes, which are also present in production systems, some answers obtained during interviews may prove to be very helpful in improving them.

The topics discussed during the interviews are listed in Tables 7.2–7.11, together with a description of the problem and a summary of the answers and comments obtained.

Based on user interviews, the following groups of issues have been identified:

- context of use
- the ability to configure the system
- gradation of the level of system warnings
- visualization of production processes
- connections between control elements and status signalling
- marginalization of some needs of operators

TABLE 7.2
Displaying Items Requiring User Intervention in the Foreground

Description of the Problem	Response of Interview Participants
If many elements require attention, indicating the gradation of their importance may be crucial for the operator's correct response. It should be noted that signalling emergency situations in the form of pop-up windows is a desirable solution in many systems; however, such a window cannot block access to other system functions. The particular complexity of production systems means that the response to an error may be so complex that it requires the operator to use functions that the system cannot associate with the emergency situation.	Some respondents suggested that since the pop-up window does not block access to other functions and can be hidden when these functions are launched, for security reasons, the pop-up window should be repeated if the problem cannot be solved.

TABLE 7.3
A Method of Marking Inactive System Functions

Description of the Problem	Response of Interview Participants
In most applications, inactive functions are shaded or hidden.	This approach was received positively, but users of production systems prefer to hide them completely. The justification here is to overload the screen with too many available functions.

TABLE 7.4
Standardization of the User Interface versus the Ability to Adapt It to the Operator's Preferences

Description of the Problem	Response of Interview Participants
In production systems, we often deal with a change of operator at the same position. Therefore, for security reasons, certain settings should be standard. However, the ability to change user interface settings affects the comfort of work and reduces the number of errors.	Users found the possibility of defining and saving individual operator settings with the option of restoring standard settings worth considering. At the same time, some users suggested that lack of time to define individual settings often prevents them from doing so. Therefore, it is worth preparing and improving predefined layouts for usability. They are especially useful when they can be the basis for defining your own settings in a short time.

TABLE 7.5
Automatic Detection of User Expectations

Description of the Problem	Response of Interview Participants
Detecting user expectations based on the frequency of using specific software functions to change its settings is now particularly common in complex applications. This allows you to highlight the most important functions.	Most users assessed the automatic change of system settings negatively. Rather, the expectation is to enable operators to make changes on their own to the greatest extent possible. Some users accept changes provided they are given the opportunity to accept them.

To sum up the interviews with employees using Information Systems for Manufacturing, it can be concluded that they are not against automating processes based on the system used, but if they are to supervise it, they must be fully informed about what action was taken and on what basis. They are against a situation where some processes take place in the background without informing the operator. This may make it difficult to detect its cause in an emergency situation. Also, configuring system settings, including changes to the interface, should be done with the consent

TABLE 7.6
Standardization of the Interface in Production Systems

Description of the Problem	Response of Interview Participants
The period when very similar monitors were used at different workstations is long gone. The general trend over the years has been to increase the resolution of displays, but their sizes have become more variable. Therefore, if you wanted to display the same content on different screens, you had to take into account the change of its layout.	Differences in the layout of program windows turned out to be not so important for users. Especially since many programs necessarily use different layouts for different screen sizes and resolutions. It is more important to use consistent command names for similar commands and icons that represent them graphically. At the same time, some users expect that predefined layouts should be available – in the form of templates – and that they will be able to save their own settings so that they can later choose a variant of working with the system.

TABLE 7.7
Language Versions of the Menu, Description and Help System

Description of the Problem	Response of Interview Participants
It happens that employees speak different languages. Current technical possibilities allow you to change the language of the entire interface. Where help is available, sometimes the base version of it is machine translated into other languages.	The interviews show that the use of the native language in the program menu is not necessary. However, it is important that the commands are defined clearly. Some users' experiences show that the translations were imprecise and even caused errors. Users have a different opinion on the language of description and help. In this regard, it is expected that the user can always choose the language for description and help. Thanks to this, he can also familiarize himself with different language versions. However, several users pointed out that it cannot be a machine translation.

TABLE 7.8
Ability to Create a List of User Own Shortcuts

Description of the Problem	Response of Interview Participants
Many systems have the ability to add keyboard shortcuts and it is common to say that this is positively received by users.	Users of production systems did not confirm in the interview that the ability to add shortcuts is always positively received. They agreed that this could be the cause of errors. Therefore, abbreviations should be standardized, and employees should be informed about modifications introduced in this area.

TABLE 7.9
Simultaneous Control of Similar Processes in the System to Improve Work Efficiency

Description of the Problem	Response of Interview Participants
Manufacturing systems control many processes. Since each process must be started and supervised by operators, combining processes into groups can improve the work of operators. The question about the possibility of integrating control of multiple functions in production systems turned out to be ambiguous for the operators of these systems. However, in the answers, it turned out to be crucial to distinguish whether integration is to be linked to the production process.	The interviews show that users create a kind of scenarios for various complex operations performed in the system. Combining these scenarios with the technological process is intended to prevent errors. Therefore, combining the control of similar processes in the system in order to improve work can take place when there is no risk that an operation in the technological process will be incorrectly indicated. It should also be remembered that this type of threat itself is an additional psychological burden for operators.

TABLE 7.10
Cooperation with the Internet of Things (IoT)

Description of the Problem	Response of Interview Participants
In the case of direct connection of sensors monitoring the production process to the production system, automatic correction of production parameters (e.g., material processing parameters) and signalling of emergency states requiring operator intervention are assumed. It is possible to link the signalling of emergency states with stopping the production process.	Such solutions are not that new, and users already have an opinion on such automation of production control. Employees' expectations mainly concern a clearly defined scope of changes in production parameters and signalling which sensor influenced the change in production parameters or signalled a critical failure. This is intended to detect incorrect sensor readings and possibly replace them quickly. It is also easier to decide to ignore the readings of one of the sensors.

TABLE 7.11
The Use of Artificial Intelligence in Production Control

Description of the Problem	Response of Interview Participants
Artificial intelligence can be used to interpret data in the production process, including readings from sensors monitoring the technological process. In this case, redundancy of the sensors used is useful so that erroneous data can be eliminated. System "learning" may be based on the variability of production process parameters and operators' decisions in specific system states. This may consequently allow for an increase in the speed of response to individual data from the production process and even reduce the number of people supervising the technological process.	In the case of systems supported in decision-making by AI, users primarily expect information about what decisions were made and what was their basis. The operator's reaction depends on this, especially when the course of the production process is questionable. Users also expect the ability to quickly report parameters of unusual system operation.

of these employees. This also applies to the available language versions and the verification of the unambiguity of the terms used in different versions.

7.5 SUMMARY AND CONCLUSIONS

Information Systems for Manufacturing are among the most complex information systems and are usually characterized by high dynamics of change. Errors during production control may result in a threat to employees and high financial losses in the event of a failure. In these systems, particular emphasis should be placed on adapting them to the psychophysical capabilities of the operators. Users have confirmed that the complexity of these systems is a major obstacle for people who are just starting to work with these systems. Since you can't afford to make mistakes due to inexperienced employees, special emphasis is placed on better training than with other software. However, training cannot always prepare for real tasks as an operator. Equipping real systems with a simulation mode will allow not only to improve the skills of operators, but also to detect imperfections of the systems themselves in terms of adapting them to humans.

IT systems used for production control are designed essentially similarly to other types of software, if we take into account those elements that are directly associated with the human–computer relationship: user communication with the system, user interface, system messages, system response time and help system. Moreover, Information Systems for Manufacturing are becoming more and more complex, among other things, because production systems use more and more complex devices, which are usually equipped with more and more complex mechanisms that monitor their operation and transmit this data to the production supervision system. Unlike new-user-oriented software, this is dominated by users who have been using it for years, many hours a day. In such a case, all requirements related to system configurability and adaptation to user needs are of fundamental importance. However, any decisions in this regard should take into account employees' opinions. What did employees expect from the interviews?

It can be concluded that the opinions of employees using Information Systems for Manufacturing indicate a very responsible approach to the possibility of improving these systems. In each case, the risk of operator's error is analysed. This does not mean that they are against further automation of decision-making processes or the use of AI. However, this should be done with full transparency of the decision-making process, so that by analysing even a failure in the system, it is possible to trace what led to it and prevent a similar situation in the future.

Since the assessment of ergonomic features concerns the interaction between a person and a system, the result of this assessment can be a wide range of outcomes. User requirements are different and change over time. This is confirmed by the varied answers obtained during the interviews. They are a spectrum of perceptual possibilities, previous experiences and expectations of individual people. Some expectations may be different and some may be difficult to meet.

The research shows that the complex context of use that occurs in production systems makes the work of operators somewhat burdensome. Trying to adapt the device and system to the user should not mean adapting it to average expectations, but rather to the expectations of a given user. Interviews show that better adaptation to ergonomic requirements involves taking into account individual user assessments. The current practice was to adapt the workstation to the average user. This was due to hardware limitations, the inability to differentiate workstations operated by many employees, or companies cutting costs. However, with software, the customization possibilities seem much greater and often less expensive. This is confirmed by the analysis of the interviews conducted. It is understandable that convergent statements show the direction of change. However, many divergent user expectations can be met by allowing the system to be configured more freely, and in some cases, by simply not blocking the ability to change certain system settings. It can be concluded that an individual approach to improving workstations may be the next stage in improving their ergonomic features.

REFERENCES

Bevan, N., 2001, International standards for HCI and usability, *International Journal of Human-Computer Studies*, 55, 533–552.

Bratukhin, A., and Sauter, T., 2011, Functional analysis of manufacturing execution system distribution, *IEEE Transactions on Industrial Informatics*, 7(4), 740–749.

Brooke, J., 2013, SUS-retrospective, *Journal of Usability Studies,* 8(2), 29–40.

Chewning, E.C. Jr., and Harrell, A.M., 1990, The effect of information load on decision makers' cue utilization levels and decision quality in a financial distress decision task, *Accounting, Organizations, and Society*, 15, 527–542.

Eppler, M.J., and Mengis, J., 2004. The concept of information overload: A review of literature from organization science, accounting, marketing, MIS, and related disciplines, *The Information Society*, 20, 325–344.

Garrett, J. J., 2011, *The Elements of User Experience: User-Centered Design for the Web and Beyond.* Berkeley, CA: New Riders.

Hankiewicz, K., 2012, Ergonomic characteristic of software for enterprise management systems. In: V. Peter (ed.), *Advances in Social and Organizational Factors.* Boca Raton, FL: CRC Press, pp. 279–287.

Hankiewicz, K., and Jayathilaka, K.R.K., 2018, Usability of an open ERP system in a manufacturing company: An ergonomic perspective. In: P.M. Arezes, J.S. Baptista, M.P. Barroso, P. Carneiro, P. Cordeiro, N. Costa, R.B. Melo, A.S. Miguel, and G. Perestrelo (eds.), *Occupational Safety and Hygiene VI.* London: Taylor & Francis Group and CRC Press, pp. 471–476.

Hankiewicz, K., and Prussak, W., 2005, Usability estimation of quality management system software. In: G. Salvendy (ed.), *HCI International. 11th International Conference on Human-Computer Interaction, vol. 4: Theories, Models and Processes in HCI.* Las Vegas NV: MIRA Digital Publ.

Hemp, P., 2009, Death by information overload. *Harvard Business Review*, 2009, 83–89. https://books.google.com/ngrams/graph?content=User+Experience. Accessed 12 May 2023.

ISO 9241-11, 2018, *Ergonomics of Human-System Interaction - Part 11: Usability: Definitions and Concepts.* Geneva: International Organization for Standarization.

ISO 9241-110, 2020, *Ergonomics of Human-System Interaction - Part 110: Interaction Principles.* Geneva: International Organization for Standarization.

ISO 9241-210, 2019, *Ergonomics of Human-System Interaction - Part 210: Human-Centred Design for Interactive Systems.* Geneva: International Organization for Standarization.

Krug, S., 2006, *Don't Make Me Think! A Common Sense Approach to Web Usability*, Second Edition. Indianapolis, IN: New Riders.

Nielsen, J., 1993, *Usability Engineering*. London: Academic Press.
Nielsen, J., 2012, Usability 101: Introduction to usability. https://www.nngroup.com/articles/usability-101-introduction-to-usability/ Nielsen Norman Group. Accessed 7 June 2021.
Norman, D., and Nielsen, J., 2022, The Definition of User Experience (UX). https://www.nngroup.com/articles/definition-user-experience. Accessed 9 June 2022.
Perret, V., Stanton, N.A., Bach, C., Calvet, G., and Chevalier, A., 2021, Validation of ergonomic criteria for the evaluation of simplex systems. *Proceedings of the 21st Congress of the International Ergonomics Association,* held online, pp. 376–383.
Pfeifer, M.R., 2021, Development of a smart manufacturing execution system architecture for SMEs: A Czech case study. *Sustainability*, 13, 10181.
Preece, J., Rogers, Y., and Sharp, H., 2002, *Interaction Design: Beyond Human-Computer Interaction*. New York: John Wiley & Sons.
Preece, J., Rogers, Y., Sharp, H., Benyon, D., Holland, S., and Carey, T., 1994, *Human-Computer Interaction*. Essex, England: Addison-Wesley Longman Limited.
Prussak, W., and Hankiewicz, K., 2007, Quality in use evaluation of business websites. In: M. Pacholski Leszek, and S. Trzcieliński (eds.), *Ergonomics in Contemporary Enterprise*. Madison, WI: IEA Press, pp. 84–91.
Sauro, J., 2011, Measuring Usability with the System Usability Scale (SUS). https://measuringu.com/sus/. Accessed 10 June 2023.
Sikorski, M., 2010, Interakcja człowiek-komputer, Wydawnictwo PJWSTK. Warszawa.
Wilson, C., 2010, *User Experience Re-Mastered Your Guide to Getting the Right Design*. Burlington, VT: Morgan Kaufmann.
Wilson, C., 2014, *Interview Techniques for UX Practitioners: A User-Centered Design Method*. Burlington, VT: Morgan Kaufmann.
Woods, D.D., Petterson, E.S., and Roth, M., 2002, Can we ever escape from data overload? A cognitive systems diagnosis, *Cognition, Technology and Work*, 4, 22–36.
Zdravković, M., Panetto, H., and Weichhart, G., 2022, *AI-enabled enterprise information systems for manufacturing, Enterprise Information Systems*, 16(5), 1–53.

8 Advanced Engineering Management Based on Intersectional R&D Challenges on Education

A Case Study for Product Classifications on Shoring Trends

Janne Heilala, Paweł Królas, and Adriano Gomes de Freitas

8.1 INTRODUCTION

Exploring the frontiers (Heilala & Krolas, 2023) of sustainable manufacturing invites us to look to the cosmos itself. Just as groundbreaking discoveries in understanding the accelerating expansion of the universe have transformed astrophysics, innovations in manufacturing exchange mechanics hold promise for a new era in industrial sustainability. This study examines how leading Finnish manufacturers and R&D operations are aligning with these shifts, drawing insights from the 2022 European Manufacturing Survey (EMS).

Using aerospace engineering as a conceptual springboard, we bridge the theoretical and the applied. Parallels emerge between mechanical engineering design and principles of spaceflight navigation. Much as Newton unpacked the forces governing the fall of an apple, this research delves into the complex factors shaping modern offshoring decisions. The EMS offers a window into Finnish manufacturing priorities, signaling how product development strategies integrate with evolving international trade frameworks like UN classification systems. Europe's progressive approach also provides guidance on how to translate sustainability from theory into practice.

As the intricate algorithm of the future of manufacturing comes into focus, Finnish industry leaders are demonstrating how to leverage innovation to drive this transformation. The confluence of scientific rigor, human-centric design, and environmental sustainability points toward a new paradigm for engineering education and industry collaboration. By exploring key issues like additive manufacturing through an

DOI: 10.1201/9781003505327-8

interdisciplinary lens, this study seeks to highlight how Finland's manufacturing sector can continue exploring sustainable frontiers (Heilala & Krolas, 2023)

8.1.1 Research Questions and Empirical Research

To cover the objectives of the study, the following research questions (RQs) were defined by selecting a case EMS:

To propel our investigation of Finnish manufacturers' approaches to sustainable global operations (Heilala & Krolas, 2023), we defined three central research questions:

- How are offshoring decisions for manufacturing and research and development guided by principles of sustainability development?

Earlier research has underscored the relevance of environmental considerations in offshore production, particularly in the case of energy efficiency infrastructure. This prompts an examination of how relocation choices account for sustainability factors on optimization.

- What protocols can guide integrated domestic and international operations to uphold sustainable priorities?

Standardized frameworks like ISO certifications provide reference points for sustainable business practices. This suggests a need to evaluate how corporations reconcile domestic and offshore protocols before delving into rabbit-holes to map the need of education technologies.

- How do quality management principles manifest in determinants of sustainable enterprise operations for aerospace design engineering example?

Establishing consistency across supply chains requires harmonizing environmental and quality standards. The research assesses how manufacturers embed sustainability into design of quality frameworks. While there is dozen way to capture and handle data, nearside is considered.

By approaching offshoring decisions through these multifaceted research questions, our analysis aims to elucidate the dynamics enabling manufacturers to integrate sustainability initiatives across local and global operations. The literature suggests that certification models could support reconciling inter/national variances, though small firms face greater obstacles in implementation. This investigation seeks to provide greater track for navigating these complexities.

8.2 EMPIRICAL LOCALIZATION

Finnish EMS dialectically offered manufacturing and R&D depthness: companies signifying offshoring and aligning into backshoring. The analysis part numbers are first withdrawn for a representative sample, and the results are elaborated with case extensions. The method introduced as advanced structure correlation modeling (Heilala & Krolas, 2023).

8.2.1 Descriptives and Interconnections

Quantitative variables derived from primary conference proceedings were first introduced. After sharpened items, the qualitative concepts were presented in a reader-friendly manner to balance the conference with the manufacturing management discussion survey items. This balanced approach investigates methodologies at the forefront of corresponding technological advancements.

The following sample characteristics withdrawn from the study were represented: minimum, maximum, mean, median, mode, standard deviation, skewness, kurtosis, sum, and validity of the sample responders on given measure indices (Heilala & Krolas, 2023, table 1, p. 233). Declension focuses on the correlation between variables emphasized by offshoring or backshoring manufacturing. The connection structure was presented accordingly. The variables in the tables take the following arguments based on abbreviations. By breaking this down, the study measure includes AT21 and AT19, representing the Annual Turnover for 2021 and 2019, respectively. NE21 and NE19 denote the Number of Employees for those same years. MCU21 and MCU19 refer to Manufacturing Capacity Utilization for 2021 and 2019, respectively. Other metrics include ROS (Return on Sales), OMP (Offshoring Manufacturing Performance), and ORD (Offshoring R&D). BFM indicates Backshoring Foreign Manufacturing. At the same time, BRD is for backshoring R&D. ET stands for Efficiency Technologies, while SDA stands for Simulation, Data Analysis, and Additive Manufacturing. The study also introduces two Energy and Efficiency Management Systems variants labeled PMC5 and PMC6. This is broken down in monodirectional declension (Heilala & Krolas, 2023, table 2, p. 234).

The sample characteristics taken from the study were represented by minimum, maximum, mean, median, mode, standard deviation, skewness, kurtosis, sum, and validity of the sample responders on the given measure indices (Heilala & Krolas, 2023, table 1, p. 233). The focus is on the correlation between variables emphasized by offshoring or backshoring manufacturing. The connection structure was presented accordingly. The variables in the tables take the following arguments based on full terms. The study's measure includes the annual turnover for 2021 and the annual turnover for 2019. Number of employees for 2021 and number of employees for 2019 are also included. Manufacturing capacity utilization for 2021 and manufacturing capacity utilization for 2019 refer to capacity. Other metrics were return on sales, offshoring manufacturing performance, and offshoring research and development. Backshoring foreign manufacturing indicates bringing manufacturing back. Backshoring research and development refers to bringing research and development back. Efficiency technologies and simulation, data analysis, and additive manufacturing were also included. The study also introduces two energy and efficiency management system variants labeled process and manufacturing control system 5 and process and manufacturing control system 6. This is broken down in monodirectional declension (Heilala & Krolas, 2023, table 2, p. 233).

8.2.2 Off- and Backshoring Manufacturing or R&D

The decision to offshore manufacturing outside of Finland comes with considerations. While there is a stringent requirement for adopting efficiency technologies, energy certifications, or environmental management systems, there is an undeniable emphasis on data analysis, simulation, and prototyping, especially using additive manufacturing. Simultaneously, the firm is expected to maintain a central hub or headquarters for research and development. Notably, from the backshoring R&D perspective, smaller companies could be incentivized to obtain these certifications (Heilala & Krolas, 2023).

Regarding smaller design offices' growth expectations without incentivizing, offshoring manufacturing enables exploration of the flexibility of the business development. Those who are operational longer and have a larger workforce show the use of certified energy management systems. This shows the balanced operational footprint in domestic regions, as noted by Heilala & Krolas (2023).

An intriguing business perspective is the financial viability of outsourcing activities to nations with a political and financial incentive for energy management and environmental conservation. According to recent research, this approach is financially sustainable, especially in countries where environmental considerations were rewarded (Heilala & Krolas, 2023). However, the taxation system often favors larger corporations, leaving smaller enterprises bereft of these benefits. Countries in Central Europe, Southwestern Europe, East and South Asia, and North America were favorable for offshoring manufacturing performance.

However, regarding research and development offshoring, Northwestern and Central Europe, the Baltics, South Asia, and the Nordic countries were preferred destinations packed with possibilities. From a logistical standpoint in nearshoring, the most sustainable relocation options for manufacturing for Finnish companies were the Nordic countries, the Baltics, and Central Europe.

8.2.3 Classifying International Trading

Human systems integration compatibility between industrial systems from legislative and economic perspectives is presented in former studies. Representation is the so-called Nomenclature Générale. Domain's integration into economic activities is selectively represented in the global schematic drawing (Heilala & Krolas, 2023, p. 236). They integratively illustrate the interplay of human and industrial systems within legislative and economic contexts. The Nomenclature Générale exemplifies this through various products, emphasizing the significance of responsibility. Quality validation, crucial for facilities such as productized universities, complements the need for product development certification in the digital era. Offshoring, with its evolving dynamics, underscores the development trajectory. The UN's ISIC (Int Classification of All Economic Activities) regional classifications emphasize understanding economic activities through global and local lenses. Discerning these by country and specific indicators for efficacious industry interpretation has implications for regional development—integrated data access, particularly in inclusive supply chain engineering (Heilala & Sajno, 2023). The progressive approach is exemplified by Eurostat standards and mirrored by the UN's interpretation. The UN's CPC

(Committee for Programme and Coordination/Central Product Classification) certified production classification (such as PRODCOM) should be a developing product taxonomy covering everything from fundamental to state-of-the-art processes. Merging classifications underlines the foreign trade statistical framework synergizing with ISIC's global trade data by managing the lookup table from (European Commission 2024; United Nations Statistics Division 2024ab). These systems reveal details about the elaboration of connections between harmonious human structures, regardless of their differences. Keeping track of these classifications is fundamental to harnessing the innovative prowess of industry leaders, meta-commanding global economic navigation, and fostering global sustainable technological integration (Heilala & Krolas, 2023).

8.3 HUMAN SYSTEMS INTEGRATION

8.3.1 Quality Standards Development

Sustainable manufacturing practices, rooted in environmental considerations, apply to technologically advanced and intensified competition (Heilala & Krolas, 2023). Smaller organizations were very flexible with the intensity of adopting these standards with the competition. Efforts from suppliers illustrate an industry-wide shift toward sustainability, extending to areas like energy efficiency and infrastructure maintenance (Gomes de Freitas et al., 2023). How, for example, can wafer optimization and training advancements improve performance and reliability within optical connections? Integrating electronic control optimization techniques requires holistic technology and skilled personnel to drive product development transformation. Finnish systems engineering education framework offers system and engineering standards guidelines, emphasizing stakeholder requirements to ensure the training models remain relevant (Heilala et al., 2023).

8.3.2 Industry Stakeholder Alignment

In continuous development, particularly in sectors like Fashion, the modern education engineering system is bound by the competitive challenge of keeping the pace (Gonzalez Chavez et al., 2022). Diamond extending business strategies, such as training simulations, offer potential solutions (Heck et al., 2010). Concurrently, as industries demand technologies for Smart Manufacturing, educational curricula must impart theoretical knowledge and practical skills to students with industry standard (Heilala et al., 2023). Sustainability challenges in global development demand a curriculum on solar energy, health protocols, and system upkeep. Though the ever-changing life sectors evolve, innovations could present opportunity for industries' research alignment throughout the organizational culture change by reverse engineering and optimizing the systems (Heilala & Sajno, 2023). Emphasizing eco-friendly practices in water and smart wafer control until the water tap prioritizes the Earth. With many solutions to climate change already in place, smart manufacturers were shifting their focus. Nevertheless, these manufacturers play a critical role in maintaining high standards across the supply chain.

8.3.3 Training for Technology Innovation Management

The specialized demands of manufacturing highlight the value of industry-focused educational content. It is essential that curricula seamlessly blend academic theory with intersectionless setting to industrial use. Thus, educating for most advanced innovation management roles is possible through science. Engineering education to technology has requirements for the synergy between structures and systems engineering. Synergizing system operations is of structural design that is not affected by intersecting factors. By default, modern technology engineering education is coupled with endless innovations. The Internet of Things archives control the current business landscape within systems (Yang, 2023). The big ascendancy of progressive e-commerce mechanisms emphasizes the urgency for businesses to grab the narrow points and connect to innovations. Design is not a word of user experience or infringes on copyright, but has seamless operations requirements for inspired design results with education. The incorporation of state-of-the-art platforms highlights examples. For example, systems applications and products courses in education that reflect the dynamic shift in business practices and strategic orientations for students to apply new management with systems, applications, and products in data processing high-performance analytic appliance.

8.3.4 Smart Integrative Solutions Are Relative

8.3.4.1 Integrating Industry 4.0

The manufacturing process's digitalization necessitates a systematic approach to ensure adaptability and efficiency. Implementing real-time feedback mechanisms in an industry setting is an educative tool for real-time rectification in tackling glitches. Manufacturers can capitalize on independent subsystems' flexibility by integrating a modular design approach, ensuring seamless updates and replacements. The audits and reviews form the cores of the system design with decoupling. This comprehensive documentation captures the design process, providing a foundation for future initiatives. At its core, the culture of innovation is recognizable from the continuous engineering of solutions supporting intersectionally accessible education adapting to the future (Figueiredo, 2022).

8.3.4.2 Infusing Traditional Curriculum

Modern engineering tools fluctuate with the dynamic demands of the manufacturing sector. As a core between theory and practice, simulation labs with the pipeline afford customers an immersive experience, while simulating real-world challenges is pedagogical in industry education (Kovalyov, 2023; Heilala & Singh, 2023a, 2023b). Peer reviews could refine collective field visit experiences to view industry practices, while continuous industry efforts for collaboration do not form silos that transdisciplinary would differentiate the industry from education. Industry collaborative initiatives for guest lectures could infuse the curriculum content for industries with rich, experiential knowledge that is completely independent of industrial protocols with continuous assessment, abandoning traditional examination for the culture of continual learning and problem-solving. Learning and motivation have a place; they remain paradoxical pitfalls between excellent learning, studying, and teaching processes.

8.3.4.3 Adapting Synchronized Setting for Education

The fusion of service and manufacturing portends a transformative shift in manufacturing in a regionally free setting until it is properly regulated. Instead of the aforementioned wafer waffle example, this integration heightens operational reasoning with black box analysis (Mirjalili & Dong, 2020). The versatility of the manufacturing service platform, such as survey, localizes in the complete management of languages and frameworks for adaptability to the research laboratories responsible for a digital ocean. Database database-service-as-a-service data protection protocols initiatives safeguard critical data sensitivity that may prove to be a complicating factor without a design involving, for example, feature aggregation in synergy with the open-ended structure. The service toolkit in manufacturing operations has initiated a manufacturing standard that is characterized by precision for innovating for enterprise resource planning systems of future. Promising manufacturing can help reveal the less visible parts of the engineering curriculum by adapting to industry trends in selective scenarios. The rise of innovative approaches to engineering curriculum aligns with the current shift toward higher quality to support industry in certification adoption.

The education sector development shows quality in systems design aligning with artificial intelligence with aerospace due to its popularity of design studies (e.g., Hoffmann Rodriguez & Elwany, 2022; Wang & Chapman, 2022). The detailed analysis of the classification of the development state to the sanctuary of the enchantress of containerization requires more design studies. The importance of advancing training and meaningful learning for an efficient control system is to adapt to another industry's requirements in order to respond to its responsible development requirements. The human–machine interaction in the software domain relates to the designs on this framework, shifting to explainable training with innovation on various platforms (Trabucchi & Buganza, 2021). Modern autonomous systems prioritizing adaptability to develop industrial facilities consider safety for example (Yu et al., 2023). Safety becomes central to the axiomatic design, which is based on independence and information axioms to address the reliable system design (Suh, 2001). Designs center on functional needs and design parameters, with system-specific frameworks for instance in aerospace. Training simulations in advanced environments can help improve physiological testing for remote sensing (Wang et al., 2022; Xu et al., 2023). Studies aim to improve system effectiveness and safety by designing simulations with physics within FPGA-based boundaries (Singh et al., 2023).

8.4 RESULTS: PRODUCT LIFE CYCLE STRATEGIES

8.4.1 Life Cycle Management Strategies for Offshoring

The analysis revealed the following key geographical preferences for offshoring different business activities:

a. Manufacturing offshoring gravitated toward Central Europe, Southwestern Europe, East/South Asia, and North America.
b. R&D offshoring preferences included Northwestern/Central Europe, South Asia, and Nordic region.

Proximity proves important, with Nordic/Baltic countries offering lower logistical hurdles.

8.4.2 Integrating Global Trade Frameworks

Classifying global economic activities related to offshoring underscores the value of international standards like ISIC. Connections emerge between regional approaches (ISIC: NAICS, NACE, ANZSIC, etc.), informing tough trade and manufacturing strategies.

8.4.3 Case Study: Sustainable Product Development

A case study of universal sustainable, innovative techniques demonstrates eco-friendly global. Energy efficiency, responsible sourcing, recyclability, and standardized components align with international benchmarks.

8.4.4 Recommendations

Key recommendations for manufacturers include:

a. Consider geographical pros/cons for offshoring specific activities
b. Leverage international classification systems customer
c. Design sustainable product lines adaptable to global
d. Participate in developing local training ecosystems

While exploratory, the analysis serves as a foundation for data-driven decision-making on globalization strategies, sustainability integration, and training innovation.

8.5 INTERPRETATION

The results show Finnish manufacturers' offshoring practices and alignment with global sustainability standards, while sustainability is of significant interest to the industry. Technological integration, in the form of simulation, data analysis, and additive manufacturing for a case product of a system, emerges, requiring manufacturing offshoring. However, the same emphasis should be observed in R&D offshoring, suggesting different strategic considerations from the design built from the global innovation hub.

As reflected in the design paradigm, training engineering for innovative transportation, incorporating energy efficiency and eco-friendly considerations, were firsthand considered globally. Environmental certifications, while valuable, are uniformly adoptable, indicating challenges for smaller entities requiring adaptation to international laws. Geographical preferences for offshoring vary, and businesses must navigate a complex global economic landscape with various classification systems.

8.6 DISCUSSION

The comprehensive data and research underscore the importance of human systems integration for understanding and predicting the behaviors of businesses considering globalization (European Commission, 2021). The drive for sustainable manufacturing practices and R&D considerations seen in the EMS data lays groundwork for future enterprises.

Cooperation between countries is intrinsically tied to policy compatibility, suggesting a need for standardized design and manufacturing practices aligned with recognized frameworks such as ISIC rather than solely regional approaches (Heilala & Krolas, 2023). The findings indicate that the curriculum for emerging technologies like containerization could align more closely with customer requirement management when the emphasis is on sustainable practices. Simulation, prototyping, and additive manufacturing require integration to ensure long-term viability and training innovation semantics according to the industry setting of time usability survey trends (Heilala & Krolas, 2023).

As the future of engineering education is tethered to sustainable digital platforms, blending containerization and international standardization forms a transformative opportunity (Cheblokov, 2023). The EMS analysis showed proximity advantages for nearshoring to Nordic manufacturing clusters. This demonstrates how the future of autonomous systems will involve innovation between advanced manufacturing techniques and localized training needs (Ojcic et al., 2023).

As exponential industrial growth introduces new complexities, equipping the next generation of engineers with the requisite knowledge and strategies in design is imperative (Panchenko et al., 2023). The EMS findings positioned training's role in risk detection and management as intertwined with sustainability initiatives. Decoupling legacy systems while nurturing new solutions will shape development trajectories (Heilala et al., 2023).

When assessing emerging autonomous technologies, continuous innovation is crucial, but potential sustainability challenges must be considered (Sinha et al., 2023). Adhering to standardized norms enhances communication reliability and safety with Certified environmental management system (ISO 14001 or EMAS) key performance indicators (Barón et al., 2022). Implementing human-centered training management elevates system intelligence, as seen in the evolution of responsible development initiatives over time (SAE International). The analysis indicates that manufacturing technology progression toward regional industrial revolutions must keep industrial metaverse safe (Mitrovic et al., 2016).

Having the initial data for global industrial simulation, this research still lacks generalization to industry, training's expanding, and technology role in risk analysis and mitigation for smart manufacturing. Mismatches in existing containerization pose threats to infrastructure, necessitating assessment updates (Mitrovic et al., 2016). Further work should explore autonomous modules for developing smarter systems. Manufacturing advancements like virtual training enhance supply chain agility. Component-level design explorations could examine climate resilience to align with Industry 5.0 aspirations, despite not being an explicit focus of past EMS analysis.

REFERENCES

A. Barón, G. Giménez, and R. Vila. EMAS environmental statements as a measuring tool in the transition of industry towards a circular economy. *Journal of Cleaner Production*, 369: 133213, 2022. https://doi.org/10.1016/j.jclepro.2022.133213.

T. Cheblokov. Advanced API performance: Pipeline state objects. NVIDIA Developer, 2023, July 18. https://developer.nvidia.com/blog/advanced-api-performance-pipeline-state-objects/.

EuropeanCommission. *Industry 5.0: Towards a sustainable, human-centric and resilient European industry*. Publications Office of the European Union, 2021. https://research-and-innovation.ec.europa.eu/knowledge-publications-tools-and-data/publications/all-publications/industry-50-towards-sustainable-human-centric-and-resilient-european-industry_en.

European Commission. (2024). PRODCOM (Production Communautaire). Eurostat. https://ec.europa.eu/eurostat/web/prodcom.

M. Figueiredo. *How to Create Artifacts in SAP HANA Cloud*, Chapter 7. Springer, 2022. https://link.springer.com/chapter/10.1007/978-1-4842-8569-5_7.

A. Gomes de Freitas, R. Borges dos Santos, L.A. Martinez Riascos, J.E. Munive-Hernandez, S. Kuang, R. Zou, and A. Yu. Experimental design and optimization of a novel solids feeder device in energy efficient pneumatic conveying systems. *Energy Reports*, 9: 387–400, 2023.

C. Gonzalez Chavez, D. Vladimirova, L. Forst, M. Despeisse, and B. Johansson. The role of trust in service-based business models the case of the fashion industry, pp. 1063–1073, 2022.

P. Heck, M. Klabbers, and M. Eekelen. A software product certification model. *Software Quality Journal*, 18: 37–55, 2010.

J. Heilala and P. Krolas. Locating a smart manufacturing based on supply chain segregation. In: V. Salminen (ed.), *Human Factors, Business Management and Society*, vol. 97 of AHFE Open Access, pp. 230–240. AHFE International, Honolulu, HI, 2023.

J. Heilala, E. Kwegyir-Afful, and J. Kantola. Training and competency development on virtual safety training. In: T. Ahram and C. Falcão (eds.), *Human Factors in Virtual Environments and Game Design. AHFE (2023) International Conference. AHFE Open Access,* vol. 96, pp. 133–143. AHFE International, Honolulu, HI, 2023. https://doi.org/10.54941/ahfe1003875.

J. Heilala and E. Sajno. Human-centric eHealth systems: A balanced design for bridging hearts and intelligent surfaces in human-oriented approach in ehealth and digital services. In: *28th Finnish National Conference on Telemedicine and eHealth, Laurea University of Applied Sciences,* October 12–13, 2023, Tikkurila, Helsinki Region, 2023.

J. Heilala, A. Shibani, and A. Gomes de Freitas. The requirements for heutagogical attunement within steam education. *International Journal of Emerging Technologies in Learning (iJET)*, 18(16): 19–35, 2023.

J. Heilala and K. Singh. Evaluation planning for artificial intelligence-based Industry 6.0 metaverse integration. In: T. Ahram, W. Karwowski, P. Di Bucchianico, R. Taiar, L. Casarotto, and P. Costa (eds.), *Intelligent Human Systems Integration (IHSI 2023): Integrating People and Intelligent Systems. AHFE (2023) International Conference. AHFE Open Access,* vol. 69, pp. 692–703. AHFE International, Honolulu, HI, 2023a. https://doi.org/10.54941/ahfe1002892.

J. Heilala and K. Singh. Sustainable human performance in large people-oriented corporations: Integration of human systems for next-generation metaverse. In: T. Ahram, W. Karwowski, P. Di Bucchianico, R. Taiar, L. Casarotto, and P. Costa (eds.), *Intelligent Human Systems Integration (IHSI 2023): Integrating People and Intelligent Systems. AHFE (2023) International Conference. AHFE Open Access,* vol. 69, pp. 386–398. AHFE International, Honolulu, HI, 2023b. https://doi.org/10.54941/ahfe1002858.

M. Hoffmann Rodriguez and A. Elwany. In-space additive manufacturing: A review. *Journal of Manufacturing Science and Engineering*, 145: 1–70, 2022.

M. Kovalyov. *Technical Potential of the Sap Hana Platform*. Litiyo i Metallurgiya (Foundry Production and Metallurgy), pp. 64–69, Belarusian National Technical University, 2023. http://dx.doi.org/10.21122/1683-6065-2023-2-64-69

S. Mirjalili and J.S. Dong (eds.). Multi-objective grey wolf optimizer. In: S. Mirjalili and J. S. Dong (eds.), *Multi-Objective Optimization Using Artificial Intelligence Techniques*. Springer Briefs in Applied Sciences and Technology. Springer, Cham, 2020. https://doi.org/10.1007/978-3-030-24835-2_5.

D. Mitrovic, M. Ivanovic, M. Vidaković, and Z. Budimac. Siebog: An enterprise-scale multiagent middleware. *Information Technology and Control*, 45, 164–174, 2016. https://doi.org/10.5755/j01.itc.45.2.12621.

Z. Ojcic, Z. Wang, and O. Litany. Sensing new frontiers with neural lidar Ffelds for autonomous vehicle simulation. NVIDIA Developer Blog, 2023, July 27. https://developer.nvidia.com/blog/sensing-new-frontiers-with-neural-lidar-fields-for-autonomous-vehicle-simulation/.

S. Panchenko, J. Gerlici, G. Vatulia, A. Lovska, A. Rybin, and O. Kravchenko. Strength assessment of an improved design of a tank container under operating conditions communications. *Scientific Letters of the University of Zilina*, 25, B186–B193, 2023.

SAE International Quality Management Systems. Requirements for aviation, space, and defense organizations (AS9100). https://www.sae.org/standards/content/ia9100/.

K. Singh, M. Saikia, K. Thiyagarajan, D. Thalakotuna, K. Esselle, and S. Kodagoda. Multi-functional reconfigurable intelligent surfaces for enhanced sensing and communication. *Sensors*, 23: 8561, 2023. https://doi.org/10.3390/s23208561.

D. Sinha, C. Shah, and A. Asif. Improve accuracy and robustness of vision AI apps with vision transformers and NVIDIA TAO. NVIDIA Developer Blog, 2023, July 25. https://developer.nvidia.com/blog/improve-accuracy-and-robustness-of-vision-ai-apps-with-vision-transformers-and-nvidia-tao/.

N.P. Suh. *Axiomatic Design: Advances and Applications*. New York: Oxford University Press, 2001.

D. Trabucchi and T. Buganza. Landlords with no lands: A systematic literature review on hybrid multi-sided platforms. *European Journal of Innovation Management*, 25(6): 64–96, 2021. https://doi.org/10.1108/EJIM-11-2020-0467

United Nations Statistics Division. (2024a). Central Product Classification (CPC). https://unstats.un.org/unsd/classifications/Econ/CPC.cshtml

United Nations Statistics Division. (2024b). International Standard Industrial Classification of All Economic Activities (ISIC). https://unstats.un.org/unsd/classifications

Y. Wang and M.P. Chapman. Risk-averse autonomous systems. *Artificial Intelligence*, 311: 103743, 2022.

Y. Wang, H. Wang, and X. Jiang. Performance of reconfigurable-intelligent-surface-assisted satellite quasi-stationary aircraft-terrestrial laser communication system. *Drones*, 6(12): 405, 2022.

Z. Xu, W. Karwowski, E. Ç akıt, L. Reineman-Jones, A. Murata, A. Aljuaid, and P. Hancock. Nonlinear dynamics of EEG responses to unmanned vehicle visual detection with different levels of task difficulty. *Applied Ergonomics*, 111: 104045, 2023.

Y. Yang. Business ecosystem model innovation based on internet of things big data. *Sustainable Energy Technologies and Assessments*, 57: 103188, 2023.

F. Yu, X. Wang, J. Li, S. Wu, J. Zhang, and Z. Zeng. Towards complex real-world safety factory inspection: A high-quality dataset for safety clothing and helmet detection, Huazhong University of Science and Technology, Wuhan, China. arXiv:2306.02098v1 [cs.CV], 1-11, 2023.

9 An Optimization Study for the Milling Process of Inconel 601 Superalloy

Yusuf Tansel Ic, Zeki Anıl Adıguzel,
Ahmed Baran Azizoglu, Alparslan Eren Keskin,
Minesu Koksal, and Berat Subutay Ozbek

9.1 INTRODUCTION

Machining processes can be classified into two types: traditional and non-traditional manufacturing processes. Since the material removal process is carried out as a result of physical contact in traditional methods, processing difficulties and other related problems may occur due to the effect of heat on the material. New superalloys have been developed in today's manufacturing industry based on the requirements of high-quality material usage. Superalloys are mainly used in the aerospace industry, considering their lightness and high-strength properties. On the other hand, machining capabilities are adversely affected in superalloys due to their high strength specifications. Machining requirements of superalloys include not only surface characteristics, but also surface hardness, electric consumption, and lead time characteristics for sustainability requirements. So, it is a challenging task to machine a superalloy with less energy and time consumption requirements while maintaining high quality levels.

Inconel 601 is a high-tech (superalloy) material widely used in industrial applications as one of the nickel alloy materials (containing approximately 60% nickel and 24% chromium). Inconel 601 contains high amounts of nickel element [1]. The main applications of Inconel 601 material include industrial heaters, heat treatment furnaces and furnace parts, exhausts, resistance heater parts, jet engines, and airline vehicles [2].

In this study, we attempted to determine the optimal values of the processing parameters (factors) by which the surface problems and energy consumption issues in the milling operation of Inconel 601 material can be minimized. For this purpose, the values of spindle speed, cutting tool feed rate, and depth of cut parameters were optimized in the milling process. Thus, quality characteristics were attempted to be optimized by determining appropriate parameter values for the milling process.

There are some studies in the literature conducted for similar purposes. In related studies, it has been seen that only one quality characteristic is concerned with optimizing the milling process of Inconel 601 material. In this study, unlike many studies

DOI: 10.1201/9781003505327-9

in the literature, multiple quality characteristics such as surface hardness, manufacturing lead time, energy consumption, and surface roughness are investigated.

In this paper, we tried to determine the most appropriate machining values for the spindle speed, depth of cut, and feed rate parameters of the four quality characteristics mentioned above in the milling operation of the Inconel 601 workpiece. The operations were made on the DOOSAN CNC machine tool (Mynx 7500/50 model) within PI Makina Company since the study aimed to respond to industrial needs.

9.2 LITERATURE SURVEY

Table 9.1 summarizes the factors (parameters) and quality characteristics (responses) used in the studies. Recent studies on the Inconel 601 process are mainly related to the machinability characteristics of the material. For example,

TABLE 9.1
Literature Summary

Reference	Parameter	Response
[3]	Cutting depth, spindle speed, feed rate	Tool life
[4]		Surface roughness
[5]		
[6]		Cutting force, surface roughness
[7]		
[8]		Surface roughness
[9]		Cutting force, power requirement
[10]		Tool life
[11]		Surface roughness
[12]		
[13]		
[14]		Cutting force, surface roughness
[15]		
[16]	Feed rate, spindle speed	Surface roughness, tool life
[17]	Cutting depth, spindle speed, feed rate	Cutting force, power requirement
[18]		Hardness
[19]		Cutting force, tool life
[20]		Surface roughness
[21]		Cutting force, flank wear of the tool
[22]		Surface roughness
[23]		
[24]		Cutting force, surface roughness
[25]	Cutting depth, spindle speed, feed	Tool life, surface roughness
[26]	rate, cutting tool costing type	Surface roughness
[27]		Burr type, surface roughness
Proposed study	Cutting depth, feed rate, spindle speed	Energy consumption, lead time, surface roughness

Korkmaz et al. [28] conducted experiments to investigate the cooling effects on machinability of Inconel 601 in dry machining, nanofluids, hybrid cooling methods, cryogenic cooling, and minimum quantity lubrication. Jovicic et al. [29] investigated the dry turning of Inconel 601 alloy using some factors such as cutting speed, feed rate, insert shapes, and machining conditions related to insert shape. They measured the flank wear, the material removal rate, and average surface roughness (Ra). They used the artificial neural network-based genetic algorithm to optimize the parameter values.

In our study, three different quality characteristics are optimized together, and our study will contribute to the literature in this respect. Another contribution of the paper is related to optimization applications. We use a weighted multi-objective optimization study to investigate different expectations from the machining process of the Inconel 601 in the parameter optimization perspective. We use different weight values scenario analysis to determine optimal parameter values based on the objective weights. So, we can obtain different optimal parameter values for cutting conditions based on the expected quality characteristics (responses) of the machining process.

9.3 EXPERIMENTAL DESIGN

Design of Experiment (DoE) enables one to find the necessary information to manage the process inputs and optimize the output. We consider controllable and uncontrollable inputs or factors in a process. We can optimize the controllable factor levels in order to improve process response using a DoE methodology. We prefer the fractional factorial experiment design method because it includes the effects of controllable factors with fewer experiment requirements. The fractional factorial experiments, or factorial designs as stated in many sources, are frequently used designs to analyze the effects of multiple parameters [30].

Three factors are determined in our study:

A. **Spindle speed (rpm)** – Low (–1) = 1115 and High (+1) = 1750,
B. **Feed rate (mm/rev)** – Low (–1) = 24 and High (+1) = 50,
C. **Depth of cut (mm)** – Low (–1) = 1 and High (+1) = 3.

In order to optimize four different responses simultaneously, in this study, the fractional factorial design was used to create an experimental design with $2^{(3-1)} = 4 \times (2)$ (replications) = 8 experiments (Table 9.2).

The Inconel 601 material was machined at the CNC machining center according to the experimental design obtained by the MINITAB R14 tool. Then, the machined surface hardness and surface roughness values were measured for each experiment (Table 9.2). Energy consumption values were calculated using experimental results. ANOVA analysis was performed using the MINITAB, and regression equations were obtained for the quality characteristics:

TABLE 9.2
Experimental Design and Results

Design of Exp.	Parameter Levels			Responses			
Experiment No.	*A* (rpm)	*B* (mm/min)	*C* (mm)	Lead Time (min) (*Y*1)	Surface Rough (Ra: μm) (*Y*2)	Surface Hardness (Brinel) (*Y*3)	Energy Cons. (kWh) (*Y*4)
1	1,115.0	50.0	1	3.13	0.280	104	2.734
2	1,7500.	24.0	1	5.27	1.10	105	4.618
3	1,115.0	24.0	3	5.21	0.830	107	4.533
4	1,750.0	24.0	1	5.34	0.640	105	4.718
5	1,115.0	50.0	3	2.59	0.420	101	2.550
6	1,750.0	50.0	1	3.13	0.310	105	2.734
7	1,115.0	50.0	3	2.39	0.270	104	2.253
8	1,750.0	24.0	3	5.53	0.440	102	5.001

$$Y1 = -5.59055\text{E}-04 \times A - 0.0969231 \times B - 0.1425 \times C + 8.742 \ (R - \text{Sq(adj)} = \%99) \quad (9.1)$$

$$Y2 = 0.000224409 \times A - 0.0166346 \times B - 0.04625 \times C + 0.922764 \ (\%46) \quad (9.2)$$

$$Y3 = -0.00086614 \times A - 0.0480769 \times B - 0.475 \times C + 107.995 \ (\%0) \quad (9.3)$$

$$Y4 = -3.40236\text{E}-04 \times A - 0.0826827 \times B - 0.05845 \times C + 7.30615 \ (\%97) \quad (9.4)$$

As a result of ANOVA analysis, it is seen that changing the factor values does not affect *Y*3. For this reason, the response optimizer tool in the MINITAB R14 tool was applied only to *Y*1, *Y*2, and *Y*4 responses.

9.4 OPTIMIZATION

A multi-response optimization was used with the regression equations of *Y*1, *Y*2, and *Y*4. The weights for each response were determined as "All Targets Equally Weighted Scenario" (Scenario 1), as shown in Figure 9.1.

According to the optimization result (Figure 9.2), the best values were determined as 1115.0 rpm, 50.0 mm/min, and 3.0 mm, respectively, for factors *A*, *B*, and *C*. Additionally, the results obtained from the counter graphs are shown in Figure 9.3.

In Figure 9.4, the minimum *Y*2 is obtained at the low level of *A* and the high level of *B*. Minimum *Y*2 was obtained at a low level of *A* and a high level of *C*. *Y*2 is minimum at the high level of *B* and both levels of *C*.

*Y*4 consumes minimum energy in *A*–*B* interaction. We can see that *A* is not very effective, but energy consumption increases as *B* decreases. If we look at *B*, there is minimum energy consumption after 47 mm/min. When parameter *A* is slower, less energy is consumed (Figure 9.5).

We developed scenario analysis using different weight scenarios for the responses (Table 9.3) and obtained new optimal parameter values based on the weights in the scenarios. The contour plots for Scenario 2, which emphasizes the lead time response

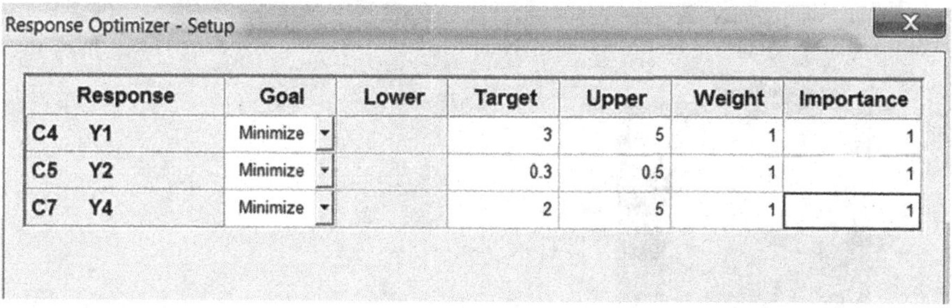

Response Optimizer - Setup

	Response	Goal	Lower	Target	Upper	Weight	Importance
C4	Y1	Minimize		3	5	1	1
C5	Y2	Minimize		0.3	0.5	1	1
C7	Y4	Minimize		2	5	1	1

FIGURE 9.1 Weight assignment for three responses in scenario 1.

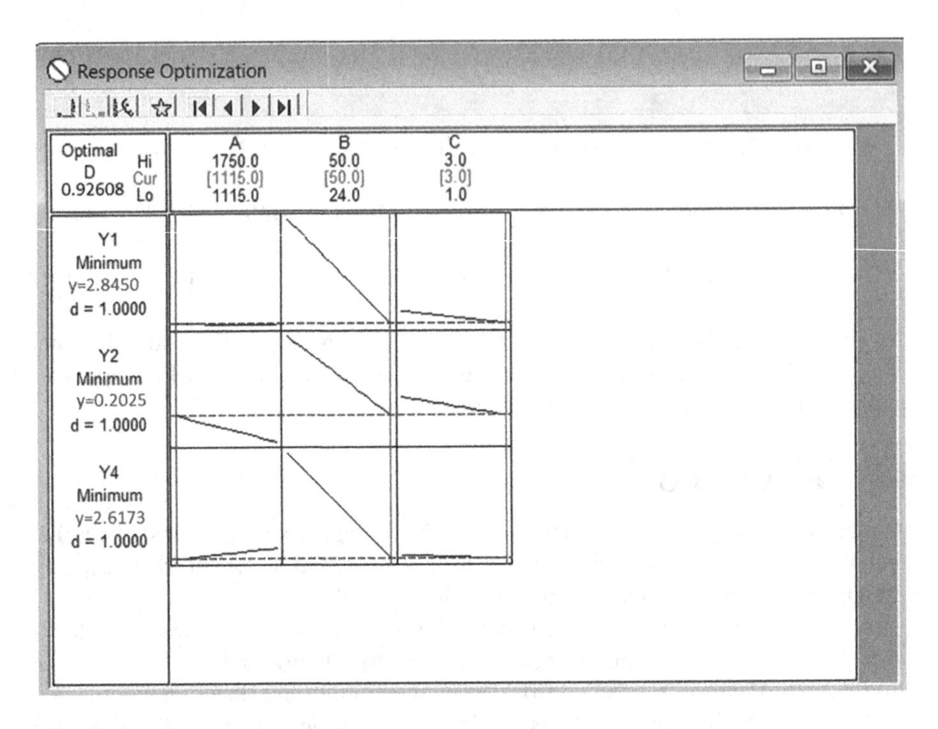

FIGURE 9.2 Response optimizer result for Scenario 1.

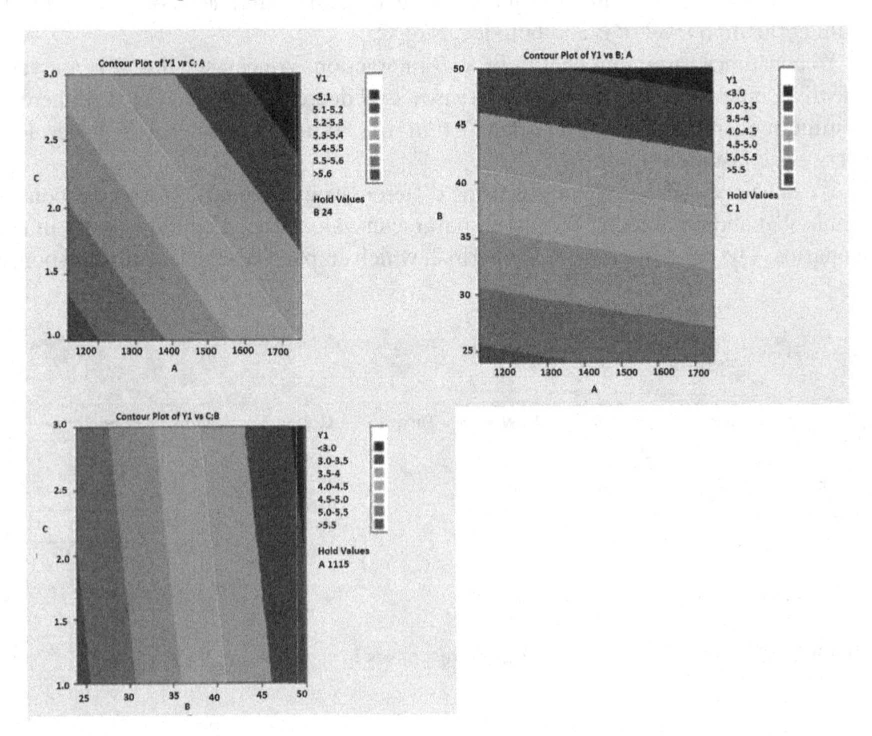

FIGURE 9.3 Response optimizer result (Scenario 1).

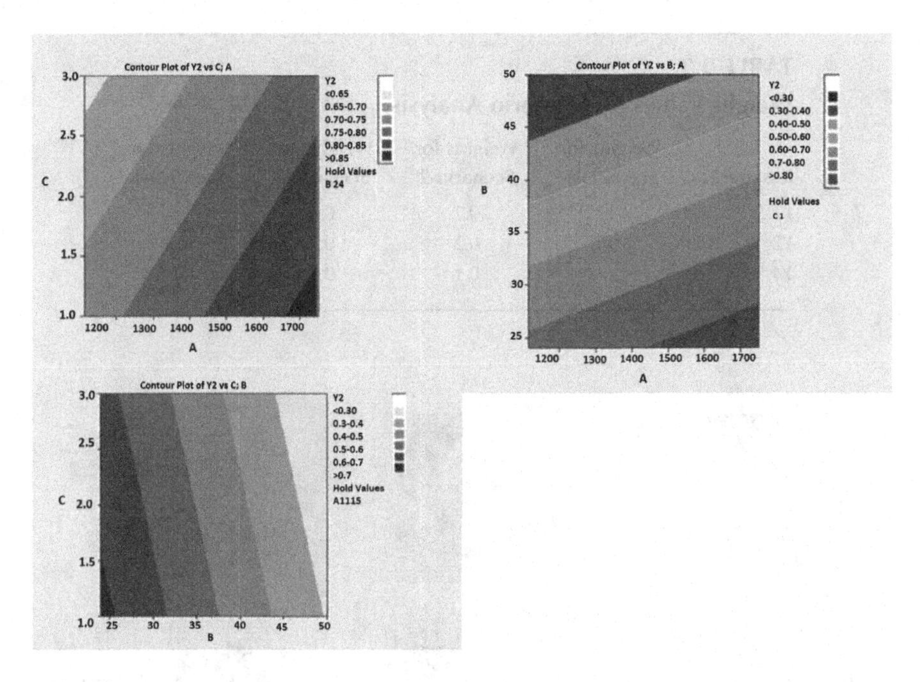

FIGURE 9.4 Contour plot analysis for *Y*2 response.

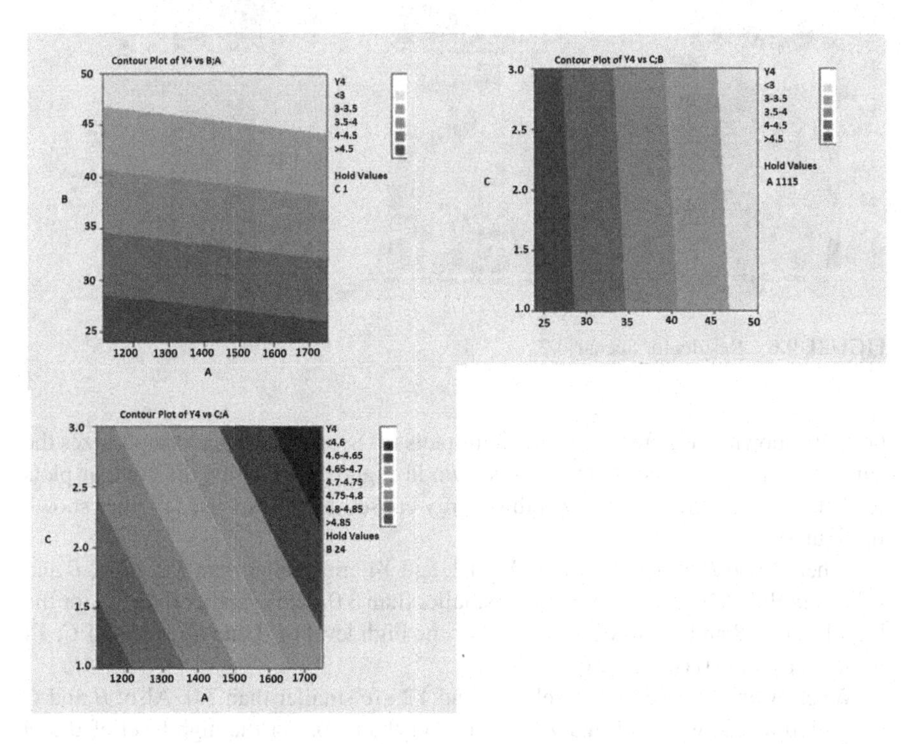

FIGURE 9.5 Contour plot for *Y*4.

TABLE 9.3
Weight Values for Scenario Analysis

Responses	Weights for Scenario-1	Weights for Scenario-2	Weights for Scenario-3	Weights for Scenario-4
*Y*1	1	0.7	0.2	0.1
*Y*2	1	0.2	0.7	0.2
*Y*4	1	0.1	0.1	0.7

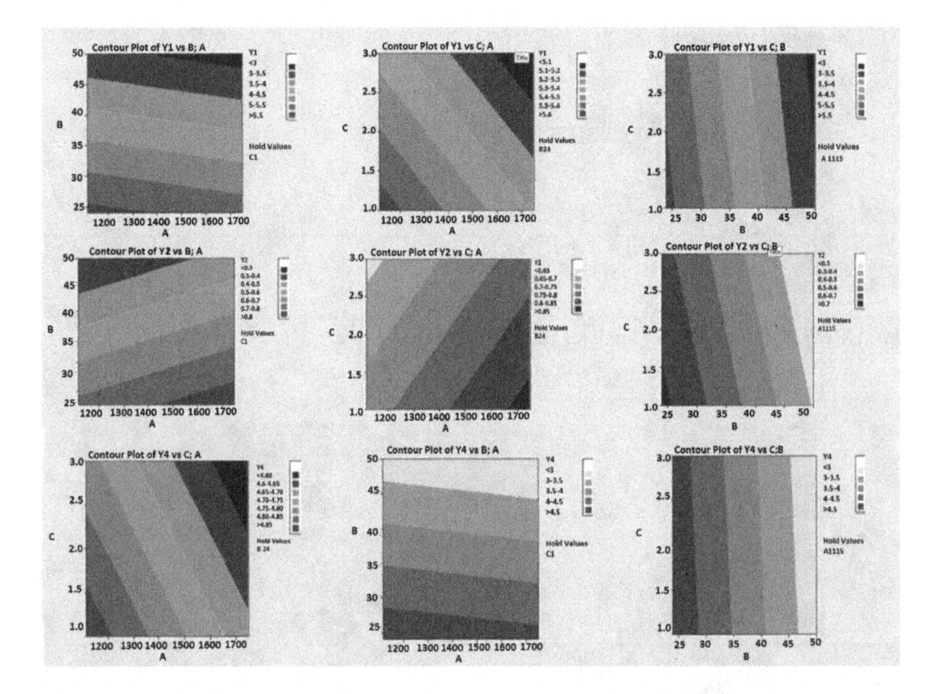

FIGURE 9.6 Results for Scenario 2.

(*Y*1), are shown in Figure 9.6. The contour plots for Scenario 3, which emphasizes the surface roughness response (*Y*2), are shown in Figure 9.7. Finally, the contour plots for Scenario 4, which emphasizes the energy consumption response (*Y*4), are shown in Figure 9.8.

When *A* and *B* have high levels, *Y*1, *Y*2, and *Y*4 are smaller than 3.0. Also, *B* and *C* have high levels, and *Y*1 and *Y*2 are smaller than 3.0. *Y*2 is smaller than 3.0 for the high level of *B* and the low level of *A*. For the high level of *B* and all levels of *C*, *Y*4 is smaller than 3.0 (Figure 9.6).

When *A* and *B* have high levels, *Y*1 and *Y*2 are smaller than 3.0. Also, *B* and *C* have high levels, while *Y*1 and *Y*2 are smaller than 3.0. For the high level of *B* and all levels of *C*, as well as the high level of *B* and all levels of *A*, *Y*4 is smaller than 3.0 (Figure 9.7).

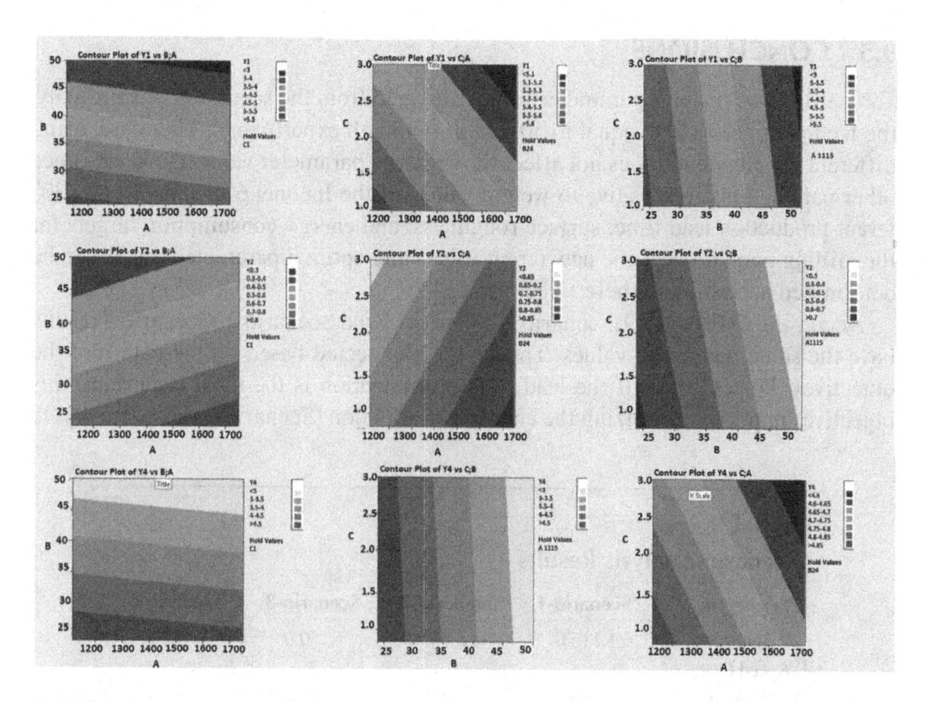

FIGURE 9.7 Results for Scenario 3.

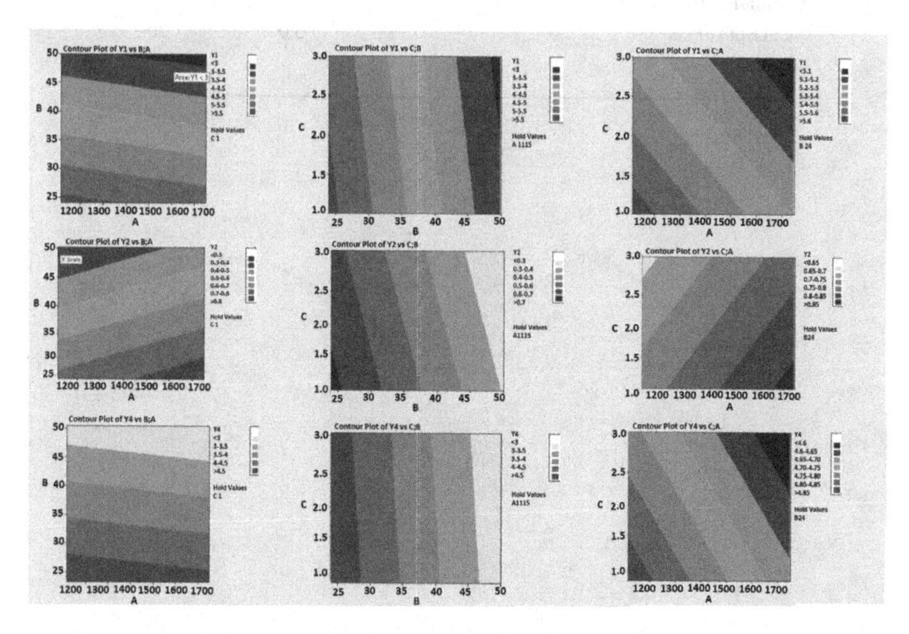

FIGURE 9.8 Results for Scenario 4.

When *A* and *B* have high levels, *Y*1 is smaller than the 3.0. Also, *B* and *C* have high levels, while *Y*1 and *Y*2 are smaller than 3.0. *Y*2 is smaller than 0.3 for the low level of *A* and the high level of *B*. For the high level of *B* and all levels of *C*, as well as the high of *B* and all levels of *A*, *Y*4 is smaller than 3.0 (Figure 9.8).

9.5 CONCLUSIONS

Table 9.4 shows the best parameter values obtained from the scenarios. Accordingly, the feed parameter (*B*) is equal to 50.0 mm/min in all experiments. Therefore, using different weight values does not affect the feed rate parameter value. However, since other parameters are sensitive to weight values, if the Inconel 601 material has different production lead time, surface roughness, and energy consumption targets for the milling operations, some new target values and optimal parameter values can be determined according to these target values.

According to Figure 9.9, Scenario 1–Scenario 3 and Scenario 2–Scenario 4 results have the same parameter values. This result is expected based on the nature of the objectives. For example, if the lead time minimization is the most important, this objective supports minimizing the energy consumption (Scenario 2–Scenario 4). On

TABLE 9.4
Scenario Analysis Results

Parameters	Scenario-1	Scenario-2	Scenario-3	Scenario-4
A: Spindle Speed (rpm)	1,115.0	1,549.50	1,150.0	1,496.5
B: Feed rate (mm/min)	50.0	50.0	50.0	50.0
C: Depth of cut (mm)	3.0	3.0	3.0	2.8

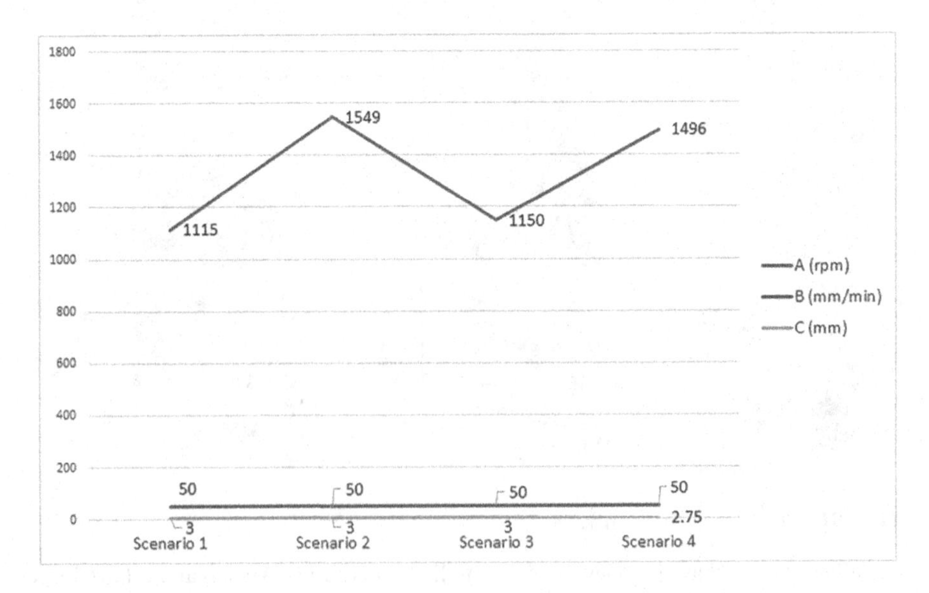

FIGURE 9.9 Graphical illustration for scenario analysis result.

the other hand, if the surface roughness (R_a) is the most important parameter, we must set a consensus between all parameters (like equal weighting scenarios: Scenario 1–Scenario 3). Another crucial result is in Scenario 4. In Scenario 4, although A is higher than Scenarios 1 and 3 results, it stabilized with a smaller depth of cut value (2.8 mm) to set a consensus between the responses' weights of energy consumption and surface roughness, 0.7 and 0.2, respectively. In this situation, a small increase of the surface roughness weight from 0.1 to 0.2 is stabilized by a high spindle speed (1,496.5 rpm) and a smaller depth of cut (2.8 mm). This analysis states that the weight values have a crucial effect on the optimal parameter values. So, the weighting process for the objective weighting stage is critical to the obtained optimization results.

ACKNOWLEDGMENT

The authors would like to thank PI Machinery Company for the machining opportunity of the Inconel 600 material in the facility.

DATA AVAILABILITY STATEMENT

The authors confirm that the data supporting the findings of this study are available within the book/chapter.

REFERENCES

1. Special Metals Website: chrome-extension://efaidnbmnnnibpcajpcglclefindmkaj/; https://www.specialmetals.com/documents/technical-bulletins/inconel/inconel-alloy-601.pdf (2023).
2. Defence Metal Website: Defence Metal/inconel-718, İstanbul, 19z (2020).
3. Choudhury, I. A., El-Baradie, M. A.: Machining nickel base superalloys: Inconel 718. *SAGE Journals*, 212(3), 195–206 (1998).
4. Coelho, R. T., Silva, L. R., Braghini Jr, A., Bezerra, A. A.: Some effects of cutting edge preparation and geometric modifications when turning INCONEL 718 (TM) at high cutting speeds. *Journal of Materials Processing Technology*, 148(1), 147–153 (2004).
5. Öktem, H., Özçelik, B., Kurtaran, H.: Inconel 718 in Frezelenmesi Sonucunda Oluşan Yüzey Pürüzlülüğünün Yapay Sinir Ağlarıyla Belirlenmesi. *Makine Tasarım ve İmalat Dergisi*, 7(1), 19 (2005) (In Turkish).
6. Altın, A., Gökkaya, H., Nalbant, M.: The effect of cutting speed in machine parameters on the machinability of Inconel 718 superalloys. *Journal of Engineering and Architecture of Gazi University*, 21(3), 581–586 (2006).
7. Pawade, R. S., Joshi, S. S., Brahmankar, P. K., Rahman, M.: An investigation of cutting forces and surface damage in high-speed turning of Inconel 718. *Journal of Materials Processing Technology*, 192–193, 139–146 (2007).
8. Nalbant, M., Altın, A., Gökkaya, H.: The effect of coating material and geometry of cutting tool and cutting speed on machinability properties of Inconel 718 super alloys. *Materials and Design*, 28, 1719–1724 (2007).
9. Taşliyan, A., Acarer, M., Şeker, U., Gökkaya, H., Demir, B.: The effect of cutting parameters on cutting force during the processing of Inconel 718 superalloy. *Journal of the Faculty of Engineering and Architecture of Gazi University*, 22(1), 1–5 (2007).

10. Thakur, D. G., Ramamoorthy, B., Vijayaraghavan, L.: Machinability investigation of Inconel 718 in high-speed turning. *The International Journal of Advanced Manufacturing Technology*, 45, 421–429 (2009).
11. Che Haron, C. H., Ghani, J. A., Kasim, M. S., Soon, T. K., Ibrahim, G. A., Sulaiman, M. A.: Surface integrity of Inconel 718 under MQL condition. *Advanced Materials Research*, 150–151, 1667–1672 (2011).
12. Kasim, M. S., Haron, C. C., Ghani, J. A., Sulaiman, M. A.: Prediction surface roughness in high-speed milling of Inconel 718 under MQL using RSM method. *Middle-East Journal of Scientific Research*, 13(3), 264–272 (2013).
13. Maiyara, L. M., Ramanujamb, R., Venkatesanc, K., Jeraldd, J.: Optimization of machining parameters for end milling of Inconel 718 super alloy using taguchi based grey relational analysis. *Procedia Engineering*, 64, 1276–1282 (2013).
14. Altın, A.: Optimization of turning machining parameters in Inconel 600 super alloy. *Journal of Engineering and Architecture of Gazi University*, 28(4), 677–684 (2013).
15. Amini, S., Fatemi, M. H., Atefi, R.: High speed turning of Inconel 718 using ceramic and carbide cutting tools. *Arabian Journal of Science and Engineering*, 39, 2323–2330 (2014).
16. Vinod, A. R., Srinivasa, C. K., Keshavamurthy, R., Shashikumar, P. V.: A novel technique for reducing lead-time and energy consumption in fabrication of Inconel-625 parts by laser-based metal deposition process. *Rapid Prototyping Journal*, 22(2), 269–280 (2016).
17. Kaynak, Y.: Evaluation of machining performance in cryogenic machining of Inconel 718 and comparison with dry and MQL machining. *The International Journal of Advanced Manufacturing Technology*, 72, 919–933 (2014).
18. Zhou, J., Ren, J., Tian, W.: Grey-RBF-FA method to optimize surface integrity for inclined end milling Inconel 718. *The International Journal of Advanced Manufacturing Technology*, 91(9), 2975–2993 (2017).
19. Tamang, S. K., Chandrasekaran, M., Sahoo, A. K.: Sustainable machining: An experimental investigation and optimization of machining Inconel 825 with dry and MQL approach. *Journal of the Brazilian Society of Mechanical Sciences and Engineering*, 40(8), 1–18 (2018).
20. Aytaç, A., Aztekin, K.: Investigation of the factors affecting the machinability of Inconel 718 alloy in turning by ceramic tool with Taguchi method. *Science Journal of Turkish Military Academy*, 29(2), 229–246 (2019).
21. Thrinadh, J., Mohapatra, A., Datta, S., Masanta, M.: Machining behavior of Inconel 718 superalloy: Effects of cutting speed and depth of cut. *Materials Today: Proceedings*, 26(2), 200–208 (2020). https://doi.org/10.1016/j.matpr.2019.10.128.
22. Waghmode, S. P., Dabade, U. A.: Optimization of process parameters during turning of Inconel 625. *Materials Today: Proceedings*, 19(2), 823–826 (2019).
23. Kasim, M. S., Hafiz, M. S. A., Ghani, J. A., Haron, C. H. C., Izamshah, R., Sundi, S. A., ... Othman, I. S.: Investigation of surface topology in ball nose end milling process of Inconel 718. *Wear*, 426–427, 1318–1326 (2019).
24. Pereira, W. H., Delijaicov, S.: Surface integrity of Inconel 718 turned under cryogenic conditions at high cutting speeds. *The International Journal of Advanced Manufacturing Technology*, 104, 2163–2177 (2019).
25. Thirumalai, R., Seenivasan, M., Panneerselvam, K.: Experimental investigation and multi response optimization of turning process parameters for Inconel 718 using TOPSIS approach. *Materials Today: Proceedings*, 45, 467–472 (2021).
26. Ji, H., Gupta, M. K., Song, Q., Cai, W., Zheng, T., Zhao, Y., ... Pimenov, D. Y.: Microstructure and machinability evaluation in micro milling of selective laser melted Inconel 718 alloy. *Journal of Materials Research and Technology*, 14, 348–362 (2021).

27. Danish, M., Aslantas, K., Hascelik, A., Rubaiee, S., Gupta, M. K., Yildirim, M. B., ... Mahfouz, A. B.: An experimental investigation on effects of cooling/lubrication conditions in micro milling of additively manufactured Inconel 718. *Tribology International*, 173, 107620 (2022).
28. Korkmaz, M. E., Gupta, M. K., Günay, M., Boy, M., Yaşar, N., Demirsöz, R., ... Abbas, Y.: Comprehensive analysis of tool wear, surface roughness and chip morphology in sustainable turning of Inconel-601 alloy. *Journal of Manufacturing Processes*, 103, 156–167 (2023).
29. Jovicic, G., Milosevic, A., Kanovic, Z., Sokac, M., Simunovic, G., Savkovic, B., Vukelic, D.: Optimization of dry turning of Inconel 601 alloy based on surface roughness, tool wear, and material removal rate. *Metals*, 13(6), 1068 (2023).
30. Montgomery, D. C.: *Design and Analysis of Experiments.* Hoboken, NJ: John Wiley & Sons (2017).

10 Empirical Study of Machine Learning for Intelligent Bearing Fault Diagnosis

Armin Moghadam and Fatemeh Davoudi Kakhki

10.1 INTRODUCTION

The manufacturing industry experiences a rapid surge in data volume due to the heightened complexity of modern manufacturing systems. Efficient utilization and modelling of such data contributes to constructing data-driven models to monitor the health conditions of manufacturing machines for improved fault detection and diagnosis.

Prognostics and Health Management (PHM) of industrial systems is a crucial part for risk assessment and reliability analysis and improvement in industrial systems [1]. PHM focuses on detecting abnormal conditions in the production process or equipment, as well as prognosis by predicting how the abnormal conditions may progress and lead to the failure in the system [1,2]. The overall framework for how PHM is useful in improving the functionality of an industrial setting as shown in Figure 10.1.

Analytical approaches that contribute to reducing downtime due to machine failure are significant in industrial settings. A crucial component in enhancing safety, reliability, performance, and overall quality of industrial processes is condition monitoring with the aim of providing insights into diagnostics and prognostics for maintenance strategical planning [3]. Creating a reliable continuous condition monitoring system with high predictivity of potential failures can be significantly useful for improving productivity in industrial systems [4].

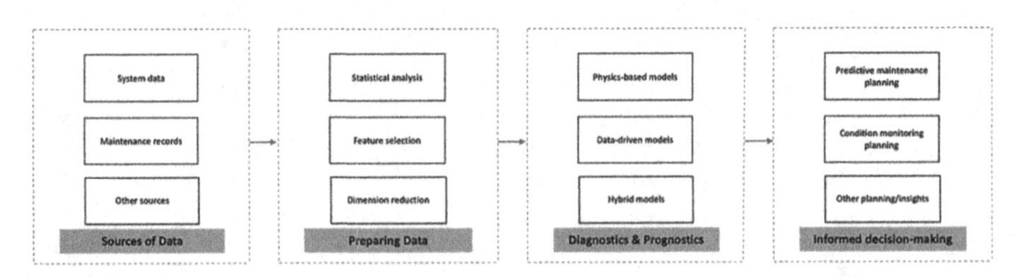

FIGURE 10.1 An overview of prognostics and health management framework for industrial systems.

DOI: 10.1201/9781003505327-10

Rolling bearings are a crucial component in rotating machinery, and efficient condition monitoring heavily relies on the effective detection of faults in rolling bearings within induction motors [5]. Bearing fault is the most frequent type of issues, accounting for 30%–40% of all the machine failures [6,7]. Hence, the importance of employing effective modelling methods for the analysis and detection of bearing faults cannot be overstated in enhancing the performance of industrial and manufacturing systems [8].

There are main approaches for detecting bearing faults. The first is model-based methods, which rely on accurate construction of failure mechanism model and mathematical methods to capture the changes of operation environments and the physical structure of the machine, which is challenging and difficult [9,10].

The other approach is data-driven modelling, which be applied for fault detection without the need for constructing a failure mechanism [11]. Traditionally, model-based approaches have been employed for fault detection, involving the creation of comprehensive system models for diagnostic and prognostic purposes. However, this method poses challenges and complexities as the system's performance being significantly impacted by variations, and the necessity for formulating specific models for each component.

An effective alternative for traditional approaches in fault diagnosis is the utilization of data-driven models [12]. A data-driven method consists of model selection and development, as well as feature extraction from raw data. The development of data-driven models involves utilizing various types of operational data sourced from a system, including but not limited to vibration data, thermal images, etc. [13]. Artificial intelligence-based machine fault analytical methods, including machine learning (ML) and deep learning, are essential for prognostics and health management of industrial systems and can be used in condition monitoring planning [14].

Publicly available datasets, such as those provided by Case Western Reserve University (CWRU), serve as valuable sources for fault detection studies. The diverse operating conditions under which data was generated enable the construction of various vibration-based analytical models. Consequently, these datasets have been extensively employed in numerous studies to assess the efficacy of models for detecting faults in induction motor bearings [15]. The abundance of operational data and the versatility of various modelling algorithms have made ML a widely adopted data-driven approach for addressing diverse regression and classification problems [16,17]. Similarly, ML techniques have found extensive application in the analysis of vibration signals, specifically for the objectives of fault detection and classification [13].

10.1.1 Machine Learning Models for Classifications

ML has been extensively utilized for constructing intricate models and algorithms to derive knowledge and provide meaningful interpretation of data. ML modelling demonstrates significant potential for inferring machine health conditions and predicting performance degradation, particularly in the era of big data [18–20]. ML algorithms have been used a popular method to better parse the data, learn from them, and apply the results of ML models for informed decision-making regarding the presence and diagnostics of bearing faults [21–23]. Various methods have been applied for detection and classification of bearing faults, specifically ML algorithms.

In classification and predictive modelling, supervised ML algorithms are employed to predict the class or category of an observation using information extracted from training and testing data points [24]. Additionally, supervised ML classification and prediction models are preferred over parametric models, as the former shows superior performance in capturing the relationship between independent variables and the target variable of interest in the analysis [24,25].

Supervised ML algorithms involve predicting target variables from a given set of independent variables through a training process [26]. These ML modelling techniques have found extensive application in the analysis of vibration signals, specifically for fault detection and classification. The choice of ML models for developing data-driven models is attributed to their capability to generalize on new data, along with their flexibility in adjusting the model structure and tuning parameters to achieve higher accuracy values.

In addition, there are many approaches for extraction of features from raw data in studying bearing faults detection and monitoring, including wavelet packet transform, time domain analysis, and frequency domain method. The time domain feature selection method can be combined with ML algorithms to distinguish between normal and faulty bearings, and it has practical application in fault detection and classification [27].

10.1.2 Goal of the Research

This study aims to assess the effectiveness of various ML classification models in detecting faults in rolling bearings within induction motors. Four distinct ML models—support vector machines, logistic regression, Naïve Bayes, and adaptive boosting decision trees—are employed for this analysis. Additionally, we examined the impact of data preparation, specifically data reduction, on enhancing the performance of ML classifiers. The evaluation focuses on the classifiers' ability to differentiate between multiple levels of faults in the rolling bearing element of the induction motor utilizing a benchmark dataset.

The rest of the paper is organized as follows: Section 10.2 discusses data used for the study and how it was prepared for modelling, followed by explanation of ML models, methodology, and experimental analyses in Section 10.3. The results and evaluation of model performance are presented in Section 10.4. The findings of the study as well as the conclusions and future research directions complete this paper are discussed in Section 10.5.

10.2 DATA SUMMARY AND TRANSFORMATION

10.2.1 Problem Statement

The purpose is to build a multi-level classifier using various ML models that is capable of accurately classifying and predicting the fault classes based on vibration data from bearings at different locations and sizes, which includes signal data generated through both faulty and normal bearing conditions.

10.2.2 Benchmark Data for Fault Detection

For this study, we utilized a benchmark dataset accessible from Case Western Reserve University (CWRU) website. The vibration data was produced and gathered using two accelerometers. Specifically, we employed a subset of the dataset generated and collected under a motor load of 1 horsepower and motor speed of 1,772 rpm, with sampling frequencies set at 48 kHz. The data was acquired from the drive end of the bearing and included a single-point fault in the bearing's inner race, outer race, and ball, each with three different sizes (0.007-, 0.014-, and 0.021-inch diameter).

The choice of input variables and the method of data preparation significantly influence the performance of ML models. In this study, we adopted a sampling approach with a gap of 2,048 readings and an overlap of 230 points to segment the dataset. Consequently, each bearing fault type comprises 230 sample points with a length of 2,048. The faults are categorized based on both size and location. The dataset includes information on three locations and three sizes for the faults, in addition to the normal bearing status data. This information was utilized to create a target variable with ten levels, where nine indicate localized faults with various sizes and one represents the normal condition. This segmented dataset serves as the initial dataset (raw data) for developing subsequent data-driven models. Appendix Figure A10.1 shows an overview of the data and the fault labels.

10.2.3 Data Preparation

Accurate and reliable automatic fault detection in practice includes eradicating the large amount of noise in the raw data through pre-processing of the data and preparing it for model development. The raw data has a substantial number of variables, demanding significant computing resources for processing. Therefore, feature extraction plays a crucial role in diminishing the dimensions of the original raw data by transforming it into more manageable subgroups, facilitating subsequent processing and model construction [28]. In this study, we employ time domain feature extraction methods to derive statistical parameters, including mean, standard deviation, kurtosis, and skewness, from signals. These parameters serve as condition features for evaluating the state of bearings. Time-domain features provide statistical information about the characteristics of the current signal and are deemed reliable features due to their sensitivities [29].

In the next phase, to reduce the data dimensions, the previously segmented data underwent transformation based on nine statistical time-domain features for each row. For every row within the segmented data, we extracted nine distinct statistics: minimum, maximum, average (mean), standard deviation, root mean square, skewness, kurtosis, crest factor, and form factor values. The mean is the average of the available data points and is used as a measure of central tendency. Standard deviation and variance are the measures of dispersion that represent the variability in the data. Standard deviation is a more precise statistical parameter compared to the range of the data, which is the difference between the maximum and minimum data points. Standard deviation is obtained by taking the square root of variance, which indicates how far data points are spread from the mean value. Skewness is another statistical parameter that represents the degree of distortion of a normal distribution and is a

TABLE 10.1
ML Model Performance on Raw Data and Transformed Data

Statistics	Equation	Statistics	Equation
Minimum	$S1 = \min(x_i)$	Root mean square	$S5 = \frac{\sqrt{\sum_{i=1}^{n} x_i^2}}{n}$
Maximum	$S2 = \max(x_i)$	Skewness	$S6 = \frac{1}{n}\frac{\sum_{i=1}^{n}(x_i - \mu)^3}{\sigma^3}$
Mean	$S3 = \frac{1}{n}\sum_{i=1}^{n} x_i$	Kurtosis	$S7 = \frac{1}{n}\frac{\sum_{i=1}^{n}(x_i - \mu)^4}{\sigma^4}$
Standard deviation	$S4 = \sqrt{\frac{\sum_{i=1}^{n}(x_i - \mu)^2}{n-1}}$	Crest factor	$S8 = \frac{x_{\max}}{x_{\min}}$
		Form factor	$S9 = \frac{\mu}{X_{\mathrm{rms}}}$

criterion for lack of symmetry in the data. Kurtosis represents the presence of outliers in the data. The crest factor represents the ratio of peak value to the effective values in the dataset. The equation on how each statistics is calculated is given in Table 10.1, where x_i represents the ith data point on each row.

This additional step notably diminishes the data's dimensions to only nine inputs for the classification of the target variable. The resulting reduced-dimension data serves as the second dataset (time domain features) for the subsequent development of ML models. Appendix Figure A10.2 shows an overview of the data and the fault labels.

10.3 MACHINE LEARNING FOR FAULT CLASSIFICATION

This section includes an overview of the ML models used in the study as well as the methodology for model development and evaluation.

10.3.1 Overview of Machine Learning Models

The description of the four ML classifiers used in the study is given as follows [30–32]:

Support vector machines (SVM) is a powerful ML modelling method that can be used for linear and nonlinear classification and regression problems, as well as outlier and anomaly detection. In classification problems, SVM models are applied to outputs that are dichotomous or multi-class. The SVM algorithm facilitates the handling of nonlinearly separable data by establishing a linear separation through the conversion of data from an input space to a feature space, employing various kernel functions.

Logistic Regression (LR) is commonly used for classification problems due to its capability in estimating the probability of an output belonging to a given class of the target variable. LR is a robust statistical technique employed for analysing multi-class output variables, utilizing a logit model. LR can compute the probability of the occurrence of a specific outcome of interest by leveraging a linear combination of the input variables.

Naïve Bayes (NB) is a simplified version of Bayesian classifiers with the assumption of conditional independence between every pair of features in the dataset [33], as well as being fast and effective in classification problems [34,35]. The NB algorithm is a component of Bayesian classifiers, employing conditional probabilities to classify an output variable based on independent input variables.

Adaptive Boosting Decision Trees (AdaBoost) is an iterative algorithm that constructs a classifier through a linear combination for the classification of a target variable. This decision tree model is constructed iteratively, utilizing a linear combination of input variables in the form of trees for classifying a target categorical variable. AdaBoost serves as a straightforward enhancement method for weak classification algorithms, effectively improving data classification by reducing both bias and variance through continuous training. The AdaBoost algorithm initiates by building distinct tree classifiers on the same training set and subsequently combines these weak classifiers to form a final strong classifier by adaptively adjusting the errors of weak learners. This process is achieved by modifying the data distribution and determining the weight of each sample based on its correct classification and the accuracy of the overall prior classification. The adjusted dataset is then fed into the subsequent classifier for training. Finally, the classifiers obtained from each training iteration are combined to produce the final decision classifier [36].

10.3.2 Development of Machine Learning Models

After preparing both datasets – one segmented and one transformed – the entire dataset was divided into training and testing sets using a 70–30 split ratio. Two sets of ML models were developed on the training data: Group 1 models were constructed using the raw segmented data, and Group 2 models were built using the extracted statistical values as inputs. The output for both groups comprised ten classes of localized faults with different sizes and normal faults. The performance of the already constructed ML classifiers in detecting and labelling data as either faulty or normal was evaluated using the 30% test data, employing the previously mentioned fault labels. The overall research methodology employed in this study is illustrated in Figure 10.2.

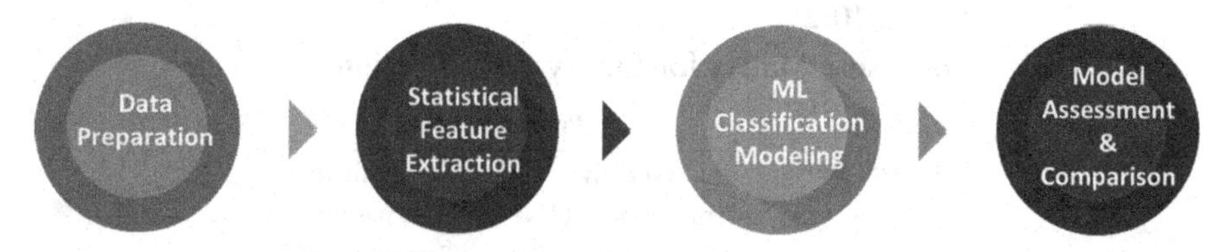

FIGURE 10.2 Research methodology used in this study.

10.3.3 Model Performance Evaluation Criteria

To compare and evaluate the performance of ML classifiers, as well as to assess the impact of data preparation and dimension reduction, two performance measurement metrics – overall accuracy and F-score – were employed. These metrics are derived from a multi-level confusion matrix generated as a result of classification, encompassing both correct and incorrect instances. For instance, if a fault in the bearing inner race with a size of 0.007 is correctly detected and classified, it is labelled as a correct instance; otherwise, it is considered incorrect. The number of accurate instances using ML predictive models compared to the observations in the original data is utilized to create a confusion matrix and compute overall model accuracy and F-score metrics. Overall accuracy values range from 0% to 100%, and F-score values range from 0 to 1, with closer values to 1 indicating a stronger predictive model in distinguishing among various levels of the target variable.

The overall accuracy rate serves as a straightforward quantitative metric for assessing the capability of an ML model classifier across all classes, without emphasizing the model's performance in classifying specific levels. It is calculated by dividing the correct number of predicted classes by the total number of cases in the dataset, as shown in Equation (10.1). Therefore, the overall error rate of the model can be directly calculated as in Equation (10.2), which is also 100% minus the overall error rate of the model.

$$\text{Overall Accuracy Rate} = \frac{TP + TN}{TP + FP + TN + FN} * 100\% \tag{10.1}$$

$$\text{Overall Error Rate} = (100\% - \text{Overall Accuracy Rate}) \tag{10.2}$$

However, the model's performance on each class is equally crucial. It indicates the model's ability to differentiate among various class output labels, shedding light on its success in accurately detecting and classifying a specific label. To accomplish this, a multi-label confusion matrix is employed. The binary classification confusion matrix is illustrated in Table 10.2. From the confusion matrix of any classifier, one can derive specificity, sensitivity, and precision, which serve to characterize the performance of a classifier for each class.

TABLE 10.2
Confusion Matrix for Binary Classification

Actual Observed Label	Predicted Label	
	Negative	Positive
Negative	True negative (TN)	False positive (FP)
Positive	False negative (FN)	True positive (TP)

Recall signifies the ratio of correctly classified labels (Equation 10.3), whereas precision assesses the concordance of the data labels with the positive labels defined by the classifier, as expressed in Equation (10.4). Another metric derived from the multi-level confusion matrix is the F-score, serving as a harmonic mean of precision and sensitivity (Equation 10.5).

$$\text{Recall} = \frac{TP}{TP + FN} \tag{10.3}$$

$$\text{Precision} = \frac{TP}{FP + TP} \tag{10.4}$$

$$F\text{-measure} = \frac{2(\text{precision} * \text{recall})}{(\text{precision} + \text{recall})} \tag{10.5}$$

10.4 RESULTS

The analysis results are presented in Table 10.3. In constructing ML models using raw segmented data, the overall accuracy fluctuates within the range of 48%–54%. Notably, LR and AdaBoost models exhibit the lowest and highest accuracy, respectively. Examining the F-score, apart from the AdaBoost classifier with a relatively low F-score of 0.513, the remaining three ML models showcase higher F-scores, ranging between 0.585 and 0.610.

When developing ML models with transformed data, utilizing time domain features as input led to a notable increase in accuracy, ranging from 91% to 96.4% across all models. This improvement was also evident in the F-score measure, with all ML models achieving values above 0.90, reaching a maximum value of 0.957 for the AdaBoost model. The results outlined in Table 10.3 underscore the effectiveness of the data transformation method employed to reduce the dimension of the training dataset, significantly enhancing the predictive power and outcomes of the models. This highlights the pivotal role of data preparation approaches in augmenting the performance of data-driven models, yielding more valuable and reliable results. As illustrated in Figure 10.3, the data transformation approach using time domain features as input led to an average increase of 42% in overall accuracy and a 0.36 improvement in the F-score.

TABLE 10.3
ML Model Performance on Raw Data and Transformed Data

Model	Accuracy-Raw Data	Accuracy-Time Domain Features	F-Score-Raw Data	F-Score-Time Domain Features
SVM	0.553	0.964	0.601	0.936
LR	0.484	0.948	0.585	0.906
NB	0.531	0.914	0.61	0.952
AdaBoost	0.542	0.976	0.513	0.957

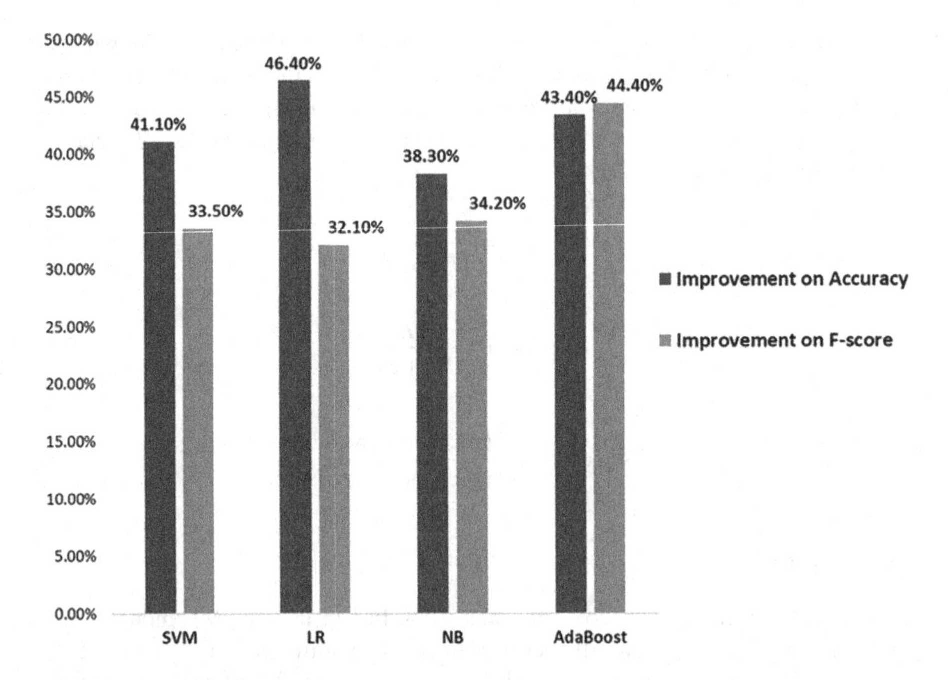

FIGURE 10.3 Improvement in ML model performance when using transformed data.

10.5 APPLICATION FOR ENHANCING MANUFACTURING EFFICIENCY

The presence of extensive data in the manufacturing sector, coupled with advancements in data storage, processing, and analysis, presents significant prospects for manufacturing intelligence and the enhancement of smart manufacturing [37]. Creating effective and dependable data-driven models capable of detecting and categorizing faults within a manufacturing system enables real-time analysis and accurate prediction of potential downtimes across the entire system. Consequently, maintenance planning strategies can be formulated based on the data-driven modules integrated into the analytics of a smart manufacturing framework.

Moreover, the adoption of data-driven models for fault detection eliminates the requirement for developing a system-specific model-based analytical approach. Unlike model-based strategies that are confined to a particular system, lack applicability to other systems, and necessitate redevelopment for distinct components, data-driven models, such as those derived from ML, exhibit generalizability. These models can be seamlessly applied to novel data from the same system, irrespective of the system model formula. The incorporation of data-driven models into the fabric of smart manufacturing not only facilitates a significant boost in the productivity and efficiency of the manufacturing sector, but also introduces a more adaptable and scalable approach to fault detection and classification. The implementation of data-driven models within smart manufacturing environments lays the groundwork for substantial improvements in manufacturing performance. Leveraging real-time data analytics for fault and anomaly detection yields multifaceted advantages that transcend traditional manufacturing practices. This includes the creation of opportunities for heightened safety protocols, enhanced operational efficiency, real-time control mechanisms, remote diagnostics, accurate prognostics, and optimized maintenance strategies. The synergy of data-driven models and smart manufacturing practices, therefore, not only addresses the limitations of traditional model-based approaches, but also paves the way for a more agile and responsive manufacturing ecosystem [38].

10.6 CONCLUSION

A prominent challenge in modelling the Case Western Reserve University (CWRU) data, and consequently obtaining reliable results, lies in the absence of a manual providing explicit instructions for classification experiments. This absence places researchers in the position to navigate the intricate task of selecting feature extraction methods, aiming to achieve not only higher accuracy rates but also meaningful and applicable models. In addressing this challenge, our study explored the analysis of the effects of data transformation and dimension reduction on enhancing the performance and reliability of ML models. This exploration specifically targets the classification of localized faults of varying sizes within the rolling bearing element of an induction motor, leveraging a benchmark dataset.

Upon comparing the performance of identical ML models on segmented raw data against models developed on time domain feature-extracted data, the latter demonstrated significantly elevated performance metrics. Notably, all models exhibited an accuracy surpassing 91% and an F-score exceeding 0.90. This noticeable improvement underscores the pivotal role of strategic data transformation and dimension reduction in bolstering the effectiveness of ML models for the precise classification of faults in rolling bearings, emphasizing their applicability in real-world scenarios.

The findings from this study underscore the crucial role of data preparation methods in shaping the overall performance of data-driven models. The accuracy and F-score values obtained are either comparable to or surpass those reported in similar studies. Moving forward, the next phase of this research involves exploring additional techniques for data dimension reduction and feature extraction. This exploration aims to advance the development of ML and deep learning models, fostering intelligent fault detection and classification based on vibration data.

ACKNOWLEDGEMENT

This publication received support from the Professional Development Grant for research dissemination, granted by the College of Engineering at San Jose State University for the author and the School of Engineering at Santa Clara University for the co-author.

REFERENCES

[1] A. Mancuso, M. Compare, A. Salo, and E. Zio, "Optimal prognostics and health management-driven inspection and maintenance strategies for industrial systems," *Reliability Engineering & System Safety*, vol. 210, p. 107536, 2021, doi:10.1016/j.ress.2021.107536.

[2] E. Zio, "Prognostics and Health Management (PHM): Where are we and where do we (need to) go in theory and practice," *Reliability Engineering & System Safety*, vol. 218, 2022, doi:10.1016/j.ress.2021.108119.

[3] J. Hendriks, P. Dumond, and D. A. Knox, "Towards better benchmarking using the CWRU bearing fault dataset," *Mechanical Systems and Signal Processing*, vol. 169, p. 108732, 2022, doi:10.1016/j.ymssp.2021.108732.

[4] A. Moghadam and F. Davoudi Kakhki, "Empirical study of machine learning for intelligent bearing fault diagnosis," In: B. Mrugalska (ed.), *Production Management and Process Control, AHFE International Conference. AHFE Open Access*, vol. 104. AHFE International, Honolulu, HI, 2023, doi:10.54941/ahfe1003049.

[5] D. Jallepalli and F. Davoudi Kakhki, "Data-driven fault classification using support vector machines," In: D. Russo, T. Ahram, W. Karwowski, G. Di Bucchianico, and R. Taiar (eds.), *Intelligent Human Systems Integration* 2021. *IHSI 2021. Advances in Intelligent Systems and Computing*, vol. 1322, 2021, doi:10.1007/978-3-030-68017-6_47.

[6] L. H. Wang, X. P. Zhao, J. X. Wu, et al. "Motor fault diagnosis based on short-time fourier transform and convolutional neural network," *Chinese Journal of Mechanical Engineering*, vol. 30, pp. 1357–1368, 2017, doi:10.1007/s10033-017-0190-5.

[7] Z. Chen, S. Deng, X. Chen, C. Li, R. V. Sanchez, and H. Qin, "Deep neural networks-based rolling bearing fault diagnosis," *Microelectronics Reliability*, vol. 75, pp. 327–333, 2017, doi:10.1016/j.microrel.2017.03.006.

[8] M. W. Milo, B. Harris, B. Bjerke, and M. Roan, "Anomaly detection in rolling element bearings via hierarchical transition matrices," *Mechanical Systems and Signal Processing*, vol. 48, no. 1–2, pp. 120–137, 2014, doi:10.1016/j.ymssp.2014.02.004.

[9] Y. Wang, D. Ning, S. Feng, "A novel capsule network based on wide convolution and multi-scale convolution for fault diagnosis," *Applied Sciences*, vol. 10, p. 3659, 2020, doi:10.3390/app10103659.

[10] G. Li, C. Deng, J. Wu, Z. Chen, and X. Xu, "Rolling bearing fault diagnosis based on wavelet packet transform and convolutional neural network," *Applied Sciences*, vol. 10, p. 770, 2020, doi:10.3390/app10030770.

[11] G. Li, C. Deng, J. Wu, Z. Chen, and X. Xu, "Rolling bearing fault diagnosis based on wavelet packet transform and convolutional neural network," *Applied Sciences (Switzerland)*, vol. 10, no. 3, 2020, doi:10.3390/app10030770.

[12] R. Zhang, Z. Peng, L. Wu, B. Yao, and Y. Guan, "Fault diagnosis from raw sensor data using deep neural networks considering temporal coherence," *Sensors (Switzerland)*, vol. 17, no. 3, p. 549, 2017, doi:10.3390/s17030549.

[13] A. Moghadam and F. Davoudi Kakhki, "Comparative study of decision tree models for bearing fault detection and classification," In: T. Ahram, W. Karwowski, P. Di Bucchianico, R. Taiar, L. Casarotto, and P. Costa (eds.), *Intelligent Human Systems Integration (IHSI 2022) Integrating People and Intelligent Systems AHFE, 2022. International Conference. AHFE Open Access*, vol. 22. AHFE International, USA Honolulu, HI. doi:10.54941/ahfe100968.

[14] S. Sun, C. Shen, and D. Wang, "Editorial for special issue: Machine health monitoring and fault diagnosis techniques," *Sensors*, vol. 23, no. 7, p. 3493, 2023, doi:10.3390/s23073493.

[15] W. A. Smith and R. B. Randall, "Rolling element bearing diagnostics using the Case Western Reserve University data: A benchmark study," *Mechanical Systems and Signal Processing*, vol. 64–65, pp. 100–131, 2015, doi:10.1016/j.ymssp.2015.04.021.

[16] S. Choppala, T. Kelmar, M. Chierichetti, and F. Davoudi, "Optimal sensor location and stress prediction on a plate using machine learning," In *AIAA SCITECH 2023 Forum*, The American Institute of Aeronautics and Astronautics, Inc., p. 0370, 2023, pp. 1–12. doi:10.2514/6.2023-0370.

[17] P. V. Badarinath, M. Chierichetti, and F. D. Kakhki, "A machine learning approach as a surrogate for a finite element analysis: Status of research and application to one dimensional systems," *Sensors*, vol. 21, no. 5, pp. 1–18, 2021, doi:10.3390/s21051654.

[18] G. E. Hinton and R. R. Salakhutdinov, "Reducing the dimensionality of data with neural networks," *Science*, vol. 313, no. 5786, pp. 504–507, 2006, doi:10.1126/science.1127647.

[19] H. Shao, H. Jiang, F. Wang, and Y. Wang, "Rolling bearing fault diagnosis using adaptive deep belief network with dual-tree complex wavelet packet," *ISA Transactions*, vol. 69, pp. 187–201, 2017, doi:10.1016/j.isatra.2017.03.017.
[20] Y. Han, B. Tang, and L. Deng, "Multi-level wavelet packet fusion in dynamic ensemble convolutional neural network for fault diagnosis," *Measurement (London)*, vol. 127, pp. 246–255, 2018, doi:10.1016/j.measurement.2018.05.098.
[21] M. A. Awadallah and M. M. Morcos, "Application of AI tools in fault diagnosis of electrical machines and drives: An overview," *IEEE Transactions on Energy Conversion*, vol. 18, no. 2, pp. 245–251, 2003, doi:10.1109/TEC.2003.811739.
[22] L. Batista, B. Badri, R. Sabourin, and M. Thomas, "A classifier fusion system for bearing fault diagnosis," *Expert Systems with Applications*, vol. 40, no. 17, pp. 6788–6797, 2013, doi:10.1016/j.eswa.2013.06.033.
[23] R. Liu, B. Yang, E. Zio, and X. Chen, "Artificial intelligence for fault diagnosis of rotating machinery: A review," *Mechanical Systems and Signal Processing*, vol. 108, pp. 33–47, 2018, doi:10.1016/j.ymssp.2018.02.016.
[24] F. Davoudi Kakhki, S. A. Freeman, and G. A. Mosher, "Evaluating machine learning performance in predicting injury severity in agribusiness industries," *Safety Science*, vol. 117, pp. 257–262, 2019, doi:10.1016/j.ssci.2019.04.026.
[25] S. S. Zhang, S. S. Zhang, B. Wang, and T. G. Habetler, "Deep learning algorithms for bearing fault diagnosticsx: A comprehensive review," *IEEE Access*, vol. 8, pp. 29857–29881, 2020, doi:10.1109/ACCESS.2020.2972859.
[26] S. E. Pandarakone, Y. Mizuno, and H. Nakamura, "A comparative study between machine learning algorithm and artificial intelligence neural network in detecting minor bearing fault of induction motors," *Energies (Basel)*, vol. 12, no. 11, p. 2105, 2019, doi:10.3390/en12112105.
[27] R. Nishat Toma and J.-M. Kim, "Bearing fault classification of induction motors using discrete wavelet transform and ensemble machine learning algorithms," *Applied Sciences*, vol. 10, no. 15, p. 5251, 2020, doi:10.3390/app10155251.
[28] R. N. Toma, A. E. Prosvirin, and J. M. Kim, "Bearing fault diagnosis of induction motors using a genetic algorithm and machine learning classifiers," *Sensors (Switzerland)*, vol. 20, no. 7, p. 1884, 2020, doi:10.3390/s20071884.
[29] M. Kang, M. R. Islam, J. Kim, J. M. Kim, and M. Pecht, "A hybrid feature selection scheme for reducing diagnostic performance deterioration caused by outliers in data-driven diagnostics," *IEEE Transactions on Industrial Electronics*, vol. 63, no. 5, pp. 3299–3310, 2016, doi:10.1109/TIE.2016.2527623.
[30] F. D. Kakhki, S. A. Freeman, and G. A. Mosher, "Use of logistic regression to identify factors influencing the post-incident state of occupational injuries in agribusiness operations," *Applied Sciences (Switzerland)*, vol. 9, no. 17, p. 3449, 2019, doi:10.3390/app9173449.
[31] F. Davoudi Kakhki, S. A. Freeman, and G. A. Mosher, "Evaluating machine learning performance in predicting injury severity in agribusiness industries," *Safety Science*, vol. 117, pp. 257–262, 2019, doi:10.1016/j.ssci.2019.04.026.
[32] W. A. Smith and R. B. Randall, "Rolling element bearing diagnostics using the Case Western Reserve University data: A benchmark study," *Mechanical Systems and Signal Processing*, vol. 64–65, pp. 100–131, 2015, doi:10.1016/j.ymssp.2015.04.021.
[33] G. Rivera, R. Florencia, V. García, A. Ruiz, and J. P. Sánchez-Solís, "News classification for identifying traffic incident points in a Spanish-speaking country: A real-world case study of class imbalance learning," *Applied Sciences (Switzerland)*, vol. 10, no. 18, p. 6253, 2020, doi:10.3390/APP10186253.
[34] T. Kumar Bhowmik, "Naive Bayes vs logistic regression: Theory, implementation and experimental validation," *Inteligencia Artificial*, vol. 18, no. 56, pp. 14–30, 2015, doi:10.4114/intartif.vol18iss56pp14-30.

[35] F. Davoudi Kakhki, S. A. Freeman, and G. A. Mosher, "Evaluating machine learning performance in predicting injury severity in agribusiness industries," *Safety Science*, vol. 117, pp. 257–262, 2019, doi:10.1016/j.ssci.2019.04.026.
[36] Y. Wu, Y. Ke, Z. Chen, S. Liang, H. Zhao, and H. Hong, "Application of alternating decision tree with AdaBoost and bagging ensembles for landslide susceptibility mapping," *Catena (Amst)*, vol. 187, p. 104396, 2020, doi:10.1016/j.catena.2019.104396.
[37] F. Tao, Q. Qi, A. Liu, and A. Kusiak, "Data-driven smart manufacturing," *Journal of Manufacturing Systems*, vol. 48, pp. 157–169, 2018, doi:10.1016/j.jmsy.2018.01.006.
[38] T. Bauernhansl, S. Kondoh, S. Kumara, L. Monostori, and B. Ka, "Manufacturing technology cyber-physical systems in manufacturing," *CIRP Annals*, vol. 65, pp. 621–641, 2016, doi:10.1016/j.cirp.2016.06.005.

APPENDIX

5	6	7	8	9	...	2039	2040	2041	2042	2043	2044	2045	2046	2047	fault
069888	0.117242	0.164389	0.200688	0.215082	...	-0.045478	-0.196098	-0.329195	-0.400750	-0.418900	-0.379263	-0.288012	-0.157087	0.012934	Ball_0.007
291853	0.188588	0.079691	-0.045687	-0.149369	...	0.118076	0.112235	0.107020	0.088662	0.060498	0.030249	0.014603	0.011891	0.010014	Ball_0.007
053823	-0.037968	-0.011057	0.023365	0.070303	...	-0.164180	-0.084906	-0.018775	0.022113	0.047982	0.065088	0.070303	0.059873	0.030249	Ball_0.007
030875	-0.038802	-0.041723	-0.037342	-0.032544	...	0.042140	0.090956	0.117033	0.127047	0.132888	0.105977	0.056118	-0.014603	-0.094503	Ball_0.007
063628	0.018775	0.105142	0.174194	0.210076	...	-0.202148	-0.180661	-0.107854	0.000626	0.109732	0.195264	0.252842	0.273078	0.261186	Ball_0.007
289767	-0.297277	-0.255137	-0.177532	-0.066757	...	0.031918	0.131010	0.191300	0.196516	0.146239	0.063210	-0.024408	-0.096798	-0.154375	Ball_0.007
065088	0.125586	0.156462	0.172942	0.168561	...	-0.225722	-0.178783	-0.113695	-0.037134	0.051528	0.135183	0.201940	0.227182	0.209033	Ball_0.007
195473	-0.221132	-0.213622	-0.176906	-0.120580	...	-0.060707	-0.061959	-0.062585	-0.048190	-0.020862	0.001878	0.019401	0.026288	0.030249	Ball_0.007
013560	0.000626	-0.001878	-0.004798	-0.017941	...	0.124961	0.162929	0.198602	0.219255	0.208407	0.157713	0.082820	0.000834	-0.070303	Ball_0.007
146657	-0.083655	-0.020862	0.043601	0.105768	...	-0.187754	-0.187545	-0.155836	-0.104308	-0.038802	0.035465	0.115990	0.184833	0.207364	Ball_0.007
191300	-0.217169	-0.214248	-0.168561	-0.083029	...	0.079691	0.094711	0.092208	0.081986	0.075519	0.072807	0.070095	0.058830	0.040054	Ball_0.007
064045	-0.085532	-0.083655	-0.055909	-0.016689	...	0.063628	0.095546	0.125795	0.159591	0.184416	0.176280	0.138938	0.076979	0.010848	Ball_0.007
149160	-0.087618	-0.018775	0.044852	0.089705	...	-0.074058	-0.106394	-0.113487	-0.105768	-0.094920	-0.078857	-0.058204	-0.018775	0.040263	Ball_0.007
132262	0.112235	0.071346	0.016481	-0.030875	...	-0.080526	-0.099510	-0.100553	-0.098258	-0.090539	-0.059038	-0.008345	0.051528	0.102847	Ball_0.007
162720	0.100553	0.029415	-0.022948	-0.060707	...	-0.002921	0.037134	0.079274	0.123292	0.171690	0.207572	0.222175	0.202983	0.155836	Ball_0.007
089079	-0.078648	-0.032961	0.020444	0.066548	...	0.009179	0.076145	0.143736	0.216334	0.273703	0.302492	0.288306	0.241368	0.181287	Ball_0.007
123500	-0.079482	-0.006258	0.073224	0.152915	...	-0.086367	-0.054657	-0.012517	0.024825	0.047356	0.049650	0.048399	0.053197	0.052154	Ball_0.007
109940	-0.125586	-0.144570	-0.158130	-0.153332	...	0.304370	0.280568	0.204652	0.091165	-0.041723	-0.161886	-0.237613	-0.280170	-0.293522	Ball_0.007
206529	0.239908	0.229894	0.205069	0.158339	...	0.277041	0.291436	0.259518	0.198810	0.129133	0.049650	-0.031501	-0.112652	-0.173151	Ball_0.007
070512	0.144779	0.221967	0.263273	0.258474	...	-0.098466	-0.129133	-0.139356	-0.127881	-0.104308	-0.085115	-0.066340	-0.046313	-0.024408	Ball_0.007
160217	-0.189214	-0.208824	-0.232606	-0.226765	...	-0.104099	-0.096589	-0.061333	-0.018984	0.019401	0.051111	0.079691	0.112026	0.128716	Ball_0.007
008345	0.024825	0.032335	0.042766	0.057369	...	0.030041	0.010639	-0.030875	-0.065505	-0.072181	-0.056743	-0.033587	0.006676	0.059664	Ball_0.007
060081	-0.002921	-0.060707	-0.099927	-0.120580	...	-0.044852	-0.077605	-0.105977	-0.121206	-0.112235	-0.082403	-0.053614	-0.028580	-0.002503	Ball_0.007
040471	-0.008136	-0.047564	-0.089079	-0.135600	...	0.077396	0.035673	0.014186	0.019610	0.043809	0.060916	0.072181	0.078231	0.091582	Ball_0.007
054657	-0.083655	-0.104099	-0.129967	-0.153332	...	-0.105977	-0.092625	-0.047564	0.008553	0.059038	0.109940	0.164389	0.219463	0.252425	Ball_0.007
029206	-0.085324	-0.129550	-0.134348	-0.100553	...	-0.196098	-0.157296	-0.102639	-0.047773	0.015855	0.084698	0.136226	0.158965	0.147908	Ball_0.007
031710	-0.053614	-0.050485	-0.032753	-0.011474	...	0.066131	0.054449	0.022739	-0.011474	-0.033378	-0.055492	-0.096172	-0.139772	-0.163554	Ball_0.007
084906	0.130176	0.170022	0.179201	0.148743	...	-0.010014	0.031918	0.090748	0.145405	0.182956	0.190049	0.165432	0.137478	0.102639	Ball_0.007
094086	-0.060290	-0.021905	0.011891	0.058621	...	-0.084072	-0.099927	-0.089705	-0.071555	-0.045895	-0.009179	0.043183	0.112235	0.174820	Ball_0.007
166058	0.118285	0.073015	0.034213	0.015438	...	0.269531	0.242411	0.170230	0.079482	-0.011891	-0.077814	-0.103265	-0.109106	-0.096675	Ball_0.007
...	...	...	...	...	...	...	...	...	...	...	...	...	...	...	...
213675	-0.269597	-0.312165	-0.327189	-0.262920	...	-0.032552	-0.037560	-0.005843	0.054253	0.128539	0.198651	0.243723	0.268763	0.278779	OR_0.021
073451	-0.151075	-0.209501	-0.251235	-0.271267	...	-0.060931	-0.132712	-0.242888	-0.340544	-0.403144	-0.348058	-0.186965	-0.050080	0.027544	OR_0.021
169437	-0.017528	-0.180288	-0.282117	-0.339709	...	0.148571	0.271267	0.358072	0.378104	0.325520	0.206163	0.065104	-0.059261	-0.161925	OR_0.021
110176	0.207832	0.256243	0.256243	0.222021	...	0.226195	0.151909	0.121027	0.136051	0.092648	0.035056	-0.024205	-0.074285	-0.070112	OR_0.021
049245	0.112680	0.146901	0.127704	0.127704	...	0.150240	0.083467	-0.009181	-0.095987	-0.164429	-0.199485	-0.214509	-0.222856	-0.214509	OR_0.021
045907	0.059261	0.066773	0.056757	0.033387	...	0.089309	0.083467	0.119357	0.079293	0.011685	-0.003339	0.017528	0.039229	0.041733	OR_0.021
180288	0.104333	0.080963	0.082632	0.110176	...	-0.050080	0.015859	0.103499	0.197816	0.272936	0.311331	0.322181	0.319677	0.291299	OR_0.021
148571	-0.129373	-0.101829	-0.060931	-0.005843	...	0.186965	0.043403	-0.093483	-0.213675	-0.285456	-0.307992	-0.309661	-0.286291	-0.226195	OR_0.021
072616	0.063435	0.039229	-0.029213	-0.116853	...	0.278779	0.252069	0.181123	0.075120	-0.038395	-0.128539	-0.189469	-0.221187	-0.218683	OR_0.021

FIGURE A10.1 Data preparation for 48K DE.

	Max	Min	Mean	SD	RMS	Skewness	Kurtosis	Crest	Form	Fault
0	0.359862	-0.418900	0.017840	0.122746	0.124006	-0.118659	-0.039330	2.901959	6.950884	Ball_0.007
1	0.467716	-0.361113	0.022255	0.132488	0.134312	0.174825	-0.078699	3.482303	6.035194	Ball_0.007
2	0.468550	-0.438092	0.020470	0.149650	0.151008	0.040369	-0.271400	3.102820	7.376930	Ball_0.007
3	0.584749	-0.543026	0.020960	0.157067	0.158422	-0.023283	0.137756	3.691090	7.558385	Ball_0.007
4	0.446854	-0.578908	0.022167	0.138189	0.139922	-0.081593	0.406102	3.193590	6.312094	Ball_0.007
5	0.437258	-0.444351	0.021119	0.138763	0.140328	-0.131424	-0.165788	3.115973	6.644547	Ball_0.007
6	0.453530	-0.491289	0.021464	0.138461	0.140082	-0.114258	0.311344	3.237607	6.526374	Ball_0.007
7	0.439553	-0.452278	0.020860	0.150120	0.151526	-0.021970	-0.269630	2.900836	7.263884	Ball_0.007
8	0.492332	-0.372170	0.020244	0.145361	0.146729	0.074225	-0.419296	3.355391	7.248032	Ball_0.007
9	0.371544	-0.490872	0.018105	0.136393	0.137556	-0.136336	-0.095047	2.701035	7.597894	Ball_0.007
10	0.387607	-0.347762	0.017720	0.126351	0.127557	-0.054709	-0.347500	3.038699	7.198568	Ball_0.007

FIGURE A10.2 Time-domain feature extracted form 48K data.

Index

Note: **Bold** page numbers refer to tables, *italic* page numbers refer to figures.